What's Really Happening on ...

Harri-Henry's FARM

Chrissy Tetley

Illustrated by Gary Young

First published in Australia in 2020 by Chrissy Tetley
Revised edition published in 2022

Copyright © Chrissy Tetley 2020

The moral right of the author has been asserted.

All rights reserved. No part of this publication may be reproduced, stored in a retrieval system or transmitted in any form or by any means without written prior permission, nor be otherwise circulated in any form of binding or cover other than that in which it is published.

All enquiries should be directed to the author.

ISBN 9780646853635 (paperback)

 A catalogue record for this book is available from the National Library of Australia

For Gilly – with love. It's never too late to have a farm!

CONTENTS

Author's Note and Disclaimer   vii

Acknowledgements   ix

What's Happening on ... Harri-Henry's Farm   1

Preface – Farming's in Your Bones – isn't it?   13

1 We Can Never Go Back to the City Now ... Where Men Wear Suits   17

2 We Celebrate Retirement   20

3 The Realistic Hobby Farmer   23

4 Farm Advice and New Best Friends   34

5 Sheep Do Funny Unexpected Things   44

6 The Drought Effect   50

7 Getting into the Nitty Gritty of Animal Management   70

8 The Farmer and Her Dogs   101

9 The Easy-Peasy Method of Caring for Cows   113

10 True Tales from the World of Living with Goats   130

11 Ponies? Come on!   144

12 How to Keep Guinea Fowl – and Why   149

13 When Mr Bond Blew Away!   167

14 Is our Farm an Efficient Pod?   173

15 The Finish Line   182

About the Author   185

About the Illustrator   186

Author's Note

Many years ago, I heard stories about an old drover and farmer called Harry-Henry. The name stayed in my mind and when I was casting around for a title for this book, I thought it would be an excellent option – thus I have used the name Harri-Henry, spelt in its feminine form, to represent my own character.

Disclaimer

I have tried to recreate events, locales and conversations from my memories of them. In order to maintain their anonymity in some instances I have changed the names of individuals and places. I may also have changed some identifying characteristics and details such as physical properties, occupations and places of residence.

Acknowledgements

I'd like to thank the many people who contributed to the creation of this book.

Bev Ryan, the Brisbane editor who started the book rolling with her early editorial advice.

Kirsty Ogden from Brisbane Self Publishing Service with her technical support to help me achieve my publishing goals. Kirsty has produced our country-style cover design and the internal page layout so that the story presents clearly.

Patrice Shaw of PS Editing who magically has a way with words.

My amazing friend, **Tina**, with her skill and patience as she photographed the many animal groups on our farm.
Pawsnap Pet Photography: www.pawsnap.com.au

Gary Young, our cartoonist, who has drawn some of the almost unbelievable true animal happenings on our farm. Gary is a cartoonist and illustrator with more than 20 years working in the newspaper industry and as a freelance artist.
www.garycartoonist.com.au

And a special thanks to my friends **Val and Bruce Bonney** of Bonnie's Dog Obedience & Care Centre (www.bonnies.com.au) for their constant encouragement, assistance and advice. Val's first book, *Who's The Boss* was the inspiration for me wanting to write an informative book about our farm animals and our hobby farm.

Finally, to all the lovely people we met as we navigated our retirement tree change and who have provided the inspiration for the characters in this book. Thank you for all the farming advice, friendship and support.

What's Happening On ... Harri-Henry's Farm

So many changes have happened on the farm since the first printing of *Harri-Henry's Farm* in 2020, that I thought it was time for an update.

We are now comfortably settled into a way of life on our hobby farm and it has been a source of wonder to me that The Engineer and I are still having so much fun, enjoying retirement in the great outdoors. Retirement? If, like me, you think there is always an adventure to be had, then this book is for you.

There is nothing glamorous about farming: I did things I told myself I'd never do, things I'll probably never forget. But there is always a laugh to be found and I like to think of myself as a kick-ass hobby farmer who has survived many funny, rewarding, hopelessly embarrassing, and downright brutal moments, managing a menagerie of farm animals, along with the challenge of the weather from wild, windy storms to the harsh droughts and heat of summer.

This lifestyle is pretty good and we've found that we love farming animals too much to quit. Yet at the same time, The Engineer and I felt we were reaching one of those crossroad moments – when we wanted to make life simpler and to slow things down to a speed that we could manage. So, when there came the opportunity for us to downsize our hobby farm from thirty acres to a mere sixteen, I told myself I would look into it. The Engineer and I had the big talk and I tried to remain rational, taking time to think about our best interests and weigh up our options.

In his practical way, The Engineer didn't take long to make a decision. 'There are great days ahead and this actually makes a lot of sense. Let's make it happen.'

So, we did. And now we enjoy farming a smaller property of sixteen acres with fewer animals.

Our New Cow Plan – It Is What It Is

The cows were first to go. Mostly I get on well with cows and our first cows, medium-sized Belted Galloways wearing snazzy white belts, were easy enough to manage. We'd bought them locally from a sensible, educated farmer, who spent considerable time handling and educating calves, and if we hadn't experienced the worst drought of all time, we would still have them. But trials and tribulations go hand in hand with life on a farm and sad things are inseparable from a farmer's experience. In the end, we ran out of ideas on how we could be more efficient getting through one of the worst droughts in history, and I had to summon up the strength to do the sensible thing: to sell off our cows.

It was a rotten summer, not even a hint of rain. Worse still, I didn't know what was to come next as the harsh drought, empty paddocks, and the awful reality of not being able to get any hay, continued week after week.

I was starting to feel more than a little anxious as I considered moving on the goats as well. We loved our goats – red and white Boer goats with bold personalities frolicking in the paddocks and nibbling on all the weeds, while producing the cutest babies every year, usually twins or triplets. In the end we had to sell the goats to another farm and this episode is still one of the saddest moments of my hobby-farming experience.

But when the summer rains did turn up, the grass and weeds grew thick and fast, so unsurprisingly, in my usual impulsive fashion, it seemed a grand idea to get more cows. Miniature cows this time, so that when the next drought threatened to disrupt our peace of mind, we wouldn't need to buy as much hay.

Not being from a practical farming background, our knowledge of cows, even little ones, was probably less than sound and things started to go wrong a few days after our five newly purchased miniature Galloways settled in. At first, I assumed that because these cows were considered miniatures and because they looked so cute and fluffy, they would be easy to manage. They *were* really cute. And they were ours. But managing them with skill was another issue. We soon realised we were in for a rough ride which came as a bit of a surprise.

Fluffy

For a while, I was almost fooled by the fluffy, white faces, but these little cows were trouble. I certainly didn't have the upper hand when the cows almost seemed to relish bouncing around us, farting, and rushing around our cattle yards in wide circles, only to narrowly skid past us and career off down the paddock.

We never knew how to assert control and soon learnt that we weren't going to herd these cows anywhere. I tried to be bold and innovative as I clutched my handy stick that seemed ineffective against the feistiness of these rollicking beasts, but instead, I ended up becoming more and more exasperated and cranky as I ran around the yards trying not to just give up and crawl back to the house. From my point of view, it was embarrassing … and I even began to find them downright scary.

The arrival of these cows in spring highlighted another problem – flies. And ticks the size of tractors. I thought 'pour-on' treatments to eliminate these pests would be straightforward enough – even for towny retirees. But I was wrong!

I began to plan a safe course of action to help us drive the cows into our new cattle yards. Our original cattle yards were huge and well into genteel decay when we bought the property, and we have since re-arranged the paddocks and rebuilt new and vastly superior cattle yards down the hill. In my plan, we would open the gates we wanted the cows to go through and reward them with some choice biscuits of lucerne hay. Then we would quickly close the slide gates and contain the cows inside the race. So, how hard could it be?

But despite me waving my arms, trying to look tall, and putting on my scary face, it didn't go to plan at all. We presumed that Fluffy, a friendly cow, would show some consideration, but every time, she kicked up a stink and shook her white fluffy head. She seemed to take much pleasure in 'not being in the mood' for a lucerne party. And then there was Biscuit, who was a force to be reckoned with. In fact, Biscuit was always a huge problem. When she kicked out with defiant enthusiasm, I soon realised it was best to run like hell and climb like a mad woman over the cattle rails. And Barbara's young calf, Thelma, was equally stubborn and it was such a performance to entice her into the race. She could have at least *tried* to remember what a relief it was *not* to have flies crawling in her eyes.

Boris

And Boris was a menace. This cute little bull, with his black nose and ears, and charming manners, who, when yarded, would try to jump the high metal rails that enclosed the race. He never could make the height and his legs would become tangled in the railings as he fell backwards. This was always a signal for Barbara and Fluffy to ricochet from the yards to the gate.

Stoically, all through two summers, we persevered, and always after a successful operation of pour-on treatments, we had huge satisfaction in knowing, 'We got them done this time!'

But enough was enough. I was spending too much energy worrying about the next time we had to 'do' a fly/tick/worm treatment and how we could best cope with getting the cows into the yards and races. A great weariness descended upon me as I realised this sort of caper wasn't good for either of us, and I really wasn't keen on having cows with a sense of humour. I kept thinking there must be easier ways for The Engineer and me to be hobby farmers and like what we do. I was over the crushing disappointment of realising that this selection of little cows was never going to work. The situation invited the question, 'Maybe it's time to sell the cows?'

Selling these pests would be a great plan! The Engineer's response was swift and clear. 'And you can continue on with your beloved sheep-breeding program without missing a beat.'

I was inclined to agree. The message was simple – The Engineer and I were not fit to be cattle farmers

This decision came at a good time. Cattle prices were high and miniature cattle were in strong demand. I was blown away by the frantic interest in our cute little cows, now that we were serious in our intention to sell.

I wasn't looking forward to meeting the new owners-to-be. It had already been one of those days and my immediate and obvious thought when they came to the farm to see the cows, was that these guys looked like *real* cattle farmers and I didn't. They were retired beef-cattle producers and straight away I decided I liked them a lot. Doug wasn't tall but he was solid, with a sunburnt face and smiling eyes, and Milly was a tiny but dominant little lady. With an air of natural authority and a relaxed manner, they considered our sweet cows grazing in the paddock. They chatted away, entertaining me with some really funny stories of their previous cattle-station experiences.

We agreed on the price for the cows as well as the portable cattle yards and features we had installed, along with the transporting of the cows the following day. Doug's preference was to load the cows into a horse float and the cattle rails onto a trailer. But I was doubtful and a little bothered; they didn't seem in anyway concerned that it might be tricky to herd five naughty, and in my opinion, feisty cows into a horse float. Especially as we had heaps of grass and so the lure of lucerne was not a high-value enticement to the cattle. But without missing a beat, Doug chuckled and nodded his head. It seemed it was his chance to really impress me.

Next morning, we waited out in the farmyard until Doug and Milly's ute and horse float came into view, their two kids driving a second vehicle and towing a massive trailer, behind them.

I simply couldn't believe my eyes when they began to work the cows. Doug was full of authority and confidence, and with respectful amusement, I watched him walk into the small yard and tweak Fluffy's tail as he pushed her up into the float. Then he waved his arms and shook his large Akubra hat to get the others to trot up behind Fluffy. It was embarrassing to see how easily the cows did what they were told. And sure enough, as soon as he pulled the ramp up and bolted it in place, five little heads poked over the top and smirked at me. I got the feeling they were quite happy to be travelling on.

And me? I was left with a feeling of reluctant relief as I waved goodbye to five annoying little cows and a set of still-newish cattle panels, all the while clutching a fistful of cash and knowing that my bank account was distinctly richer for the experience. In fact, everything was good. But this whole experience confirmed to me that despite my enthusiasm for cows, I was lacking in cattle management skills and techniques.

Despite this disappointment, I cheered myself up with the knowledge that I had enough experience and practical know-how to continue rearing our flock of Southdown sheep. Everything's going well and we're enjoying building up our mob of purebred ewes with two capable rams. And when triplets were born this year, it was a joy to watch the mama ewe nurturing and protecting her young. However, the smallest ram lamb was a concern and would not have made it without being partially hand-reared with supplementary bottles of a good-quality milk product. This treatment had a spectacular effect and it wasn't long before he rallied and was able to fend for himself. The frail, tiny triplets have against all odds grown into strong, sturdy lambs.

The Idea of Mabel

It's not just farm animals that fascinate me; I've always had a passion for dogs. Dogs are essential to my idea of what constitutes a farm family and a long time ago I fell in love with German Shepherds. I fall in love with them still and enjoy a continuously inspiring relationship with these outstanding dogs. Our dogs successfully compete in obedience trials, and the training and education we take on develops an intense focus and sheer skill at all levels of working-dog competitions.

Tomek

On the other hand, running a small flock of sheep in our great outdoors with a well-trained herding dog is fulfilling my hobby-farming dream and the theory is, that our sheepdogs are to work the sheep and sleep in the shed, but to tell the truth, they too are spoiled pets and we just love having them around.

It's extremely difficult to move a flock of sheep through our wide-open, hilly paddocks, without the help of a good sheepdog with an inbred instinct to chase sheep. When we first got Belle, I wasn't sure how to train her and it was through considerable trial and error that I ended up with a pretty good paddock partner *[see chapter 8]*.

Belle has the ability to do things I cannot do, or, if I can do them, to do them better than I can. Her natural obedience and inherent herding skills, linked to a friendly disposition, makes her invaluable on our farm herding sheep, and I decided that a new working sheepdog to run alongside seemed like a good idea. One who could serve its apprenticeship with the older, trained dog to learn some of the rudiments of herding.

This was my new fun assignment – looking for a puppy that would be willing and eager to work with me, have plenty of energy, and a natural instinct to gather sheep into a small mob. To my surprise the search proved to be difficult. Puppies for sale in our local area were either too expensive, too old, too far away to be convenient to view, or without a Border Collie heritage, which was a pre-requisite, in my opinion, for a good sheepdog. Then I came across an advert for Border Collie cross Stumpy Cattle Dog puppies in one of the local papers, and it seemed from our text exchanges that I had found just what I wanted. I organised the deposit, zoomed out the door, and headed off to a nearby commercial dairy farm to view the puppies. The sire was an eager, handsome Border Collie who leapt up at us, wagging his tail furiously and the mum was a gentle little Stumpy-style Cattle Dog. Both parents demonstrated excellent temperaments, soundness and intelligence. And all the pups sported short coats, which was a bonus, as sometimes grass seeds can cause grooming issues for dogs with long coats.

I had first pick from the litter – a mixed bag of colours, some black and white, some leaning more towards the traditional Cattle Dog colours, and one girl without a tail but with a tiny stump. And this was the pup we chose to be Mabel.

Mabel seemed perfect for our mob of sheep, our sixteen acres, and my sort of temperament!

Mabel

In my mind, Mabel was to be an indispensable working dog, but what I hadn't factored in was that I would fall in love with this tiny black-and-white, cute-as-a-button puppy. The natural obedience of this little dog was linked to her loveable disposition and, after allowing her a few days pottering around our property, it was time to introduce her to the sheep. Her confidence showed no bounds and right away she would wing around the sheep in a wide joyous gallop, her little legs pumping away as she tried to keep up.

Early training for Mabel centred around lead training and understanding how to be tethered. My commands and signals are really simple and she has adapted well to walking at heel, waiting for sheep to bound through gateways, and to not bite me. Her retrieve instinct is particularly strong and she has learnt to come when called – especially if a small piece of cheese is on offer.

Working two dogs together has some considerations, especially when there is a puppy and an older dog. Plus, we have some variations on how we herd the sheep. Sometimes I let Mabel work alone with me, sometimes she's on lead watching Belle bring the sheep into the yards, and sometimes I keep Belle on lead and give Mabel a go. And sometimes I leave Mabel in the kennels and let Belle have time on her own without our annoying, bossy, bitey puppy!

In reality, each dog must be trained individually to feel confident in her own right, to develop a good herding style – a sensible, fast but smart approach, and to accept responsibility on her own. Mabel is learning how to stop when Belle stops and to stay until told to move on. This exercise takes time and is a repetitive and reinforced action.

Our sheep recognise that Belle is in charge. But it hasn't taken Mabel long to develop the confidence to push the sheep and move them along, although she needs to learn to slow down and do this in a quiet, gentle, but masterly way.

A training book I recommend for all hobby sheep farmers is *Training and Working Dogs* (University of Queensland Press, 1987) by Scott Lithgow, a grazier and drover from our Australian outback.

What's Really Happening On ... Harri-Henry's Farm is still about our journey as hobby farmers and the animals that have inspired and shaped us, and helped us work out what we want to do in this world. Our retirement life, running animals on our hobby farm, has turned out to be one of magical fulfilment.

PREFACE

Farming's in Your Bones – Isn't it?

'How's being a farmer working out for you? Good? Keep it up then.' Old Uncle Arthur James was effortlessly cool in his summing up of our retirement plan.

Someone once said, 'Retirement is when you quit your desk job to read the papers in the morning and enjoy a cappuccino with friends in the afternoon.' But our retirement began in earnest on a sunny hillside at Clifftop Cottage Farm – and somewhat to our surprise, has turned into an extraordinary and uplifting experience. I had always liked the thought of farming animals, despite growing up in the suburbs and living in the city, and now my life wouldn't feel right without dogs and cows and sheep, and constant bursts of jumbled animal issues happening … usually all at the same time.

There is no doubt some old friends considered I was missing some marbles and took bets on how long the honeymoon would last. And, when the news was out, the family was not convinced that an escape into rural farming life would stick. The boys were momentarily taken aback: our farming lark, they thought, was slightly extreme.

'Ma! What's going on? You should both be taking it easy now Dad's retired. A farm? You're not serious, are you?'

'Not serious? Of course I'm serious!'

HARRI-HENRY'S FARM

Was I serious? I was pumped!

This book tells of how I became a farmer. My natural instinct was to only tell the good bits – the things that make me look clever and awesome. But farming isn't really like that, and sometimes the only thing to say is … @#!?&!!!

Our stories tell about the funny everyday things that happen and are, at best, gems of purest common sense. But I wanted to tell about the sad things too, because they are part of never-ending animal issues on our farm.

Were we mad or inspirational? Taking on a forty-acre wreck of a hobby farm, with all the fences in disrepair, is a bit of both. And, we couldn't have chosen a more inappropriate spot: this was about as far from the sensible environs of retirement as can be imagined. These days The Engineer, a man who still likes a challenge, is just an ordinary bloke in denim jeans, old floppy hat and sturdy work boots. A world away from the business-suit attire of the engineering hub. And I get to work at home in old jeans and faded t-shirts, play with the dogs, and still watch Dr Phil. That's seriously cool.

Preface: Farming's in Your Bones – Isn't it?

Retired friends still make comments about early morning chores and how the farm and all those animals just tie us down; the consensus being that it is too time-consuming at our 'time of life'. But, to stumble out of bed early and grab hold of farming possibilities and experiences with both hands is unbelievably magical, even when it's so toasty under the bedcovers. Our days are busy, so I never wake up and wonder what I'll do that day. As first-time farmers, we've had to tackle farm life head on, grappling with the big issues like delivering baby lambs; the drought; dealing with bloat in cows; and how to do a 'preg-test'*.

The remoteness and peace on our farm captivated us instantly. How very nice it is to shrug off the city pace and hit the country roads to home. Out here there are no street lights when darkness falls, no glittering high-rise apartments – just the spectacular stillness of the night with the whisperings of invisible nocturnal creatures. And, from every room in our house, even the bathroom, we have chocolate-box views of grassy paddocks, distant blue hills and lots of sheep. Even the flies are friendly.

And in the late afternoon, on top of our hill as the sun seemingly speeds towards the horizon, we sit with coffee in hand on old park benches and absorb the immensity of the night sky rolling in. Long bands of amber and pink rays sink into the west as distant mountains become boldly etched silhouettes against the red and orange skies. And at night the stars are unbelievable: those millions of tiny lights, beautiful, blinking, bright. This is as good as it gets and it tugs at our souls, and when wild storms sweep in after nightfall, we have a spectacular light show.

We get invited to parties, barbeques and family teas, where hardy true-blue country people tell their stories. These wonderfully earnest, warm, funny characters frequently drop by our farm for a cuppa, or engage us in conversation down at the Bull & Barrel Country Pub. And problems become less threatening and more solvable after they tell us things about animals that we didn't know we needed to know: valuable information regarding illnesses and treatments, and the basics of care when raising livestock.

Our success owes thanks to so many friends, colleagues and mates in our farming community, all prepared to share their own stories, wisdom and experiences. And special thanks to Gary, our cartoonist, for comically portraying the highlights of our retirement stories.

The final word belongs to our son Mick: 'You made it, Ma!'

Chrissy Tetley

*preg-test: Rectal palpation is the cheapest and most convenient method of pregnancy testing cattle.

ONE

We Can Never Go Back to the City Now... Where Men Wear Suits

In the beginning, I was quite happy when The Engineer retired. I was loving the luxury of indulging in long sleep-ins and having someone pottering around the house, sharing my every whim and want. For some time, The Engineer had been looking forward to handing over his desk job in the tall grey cubicle-floored office building, but then retirement began to present a problem as we worried about what sort of life people our age should be having.

Retirement is a defining, life-changing, curious thing: I get that. You can't be retired for long without thinking of other plans down the track, and it's often taken for granted that oldies, like us, shouldn't care too much if they have nothing much to do. A profound depression descended upon me as my mind kept drifting helplessly over the mundane choices we were thinking about, like buying a really nice house in the suburbs, close to the railway station, a bus stop and a supermarket. Quite simply, this thought was unthinkable, and I was rapidly coming to the conclusion that retirement was just a big empty life-space looming ahead of me: one huge disappointment with no obvious appealing features.

I realised I was bored and I yearned to do something new, something quite out of the ordinary, and I was looking for something I could do about it. Friends were not shy with suggestions: 'Go on a cruise; they're amazing. Spend some lazy days touring on a boat to Europe. Maybe take up a new hobby – golf, knitting, grow vegetables and roses – or play bridge.'

Yet, incredibly it was The Engineer who first raised the possibility. 'When you think about it, we need a tree change. How good would it be to buy a farm?'

That did it. I was ecstatic: why hadn't I thought of this before? The whole idea was overwhelming. It was huge. I was soon imagining me on a gorgeous farm with fat animals and lots of dogs, squeaky-clean barns, and lush green paddocks.

His suggestion to buy a farm was a life-changing moment and my enthusiasm showed no bounds, as I began a relentless search online. Looking for the perfect property did take some consideration as it was to be a significant investment, both financially and emotionally. I looked everywhere for something that we liked and could afford as retirees; but as time went by, my search began to be somewhat depressing.

I was beginning to think maybe I was looking in the wrong places. Then one evening I came across a worn-out rural property known as Clifftop Cottage Farm. The location was perfect, the price negotiable, and my heart began to thump out of control.

There was a lot hanging on our first visit to Clifftop Cottage Farm. We met up with the real estate agent for a site inspection, and drove most of the way on a dusty road winding over hills, to look through a farm cottage on forty acres with big sky views. To me it felt just perfect and kind of hard to believe. Even so, The Engineer was apprehensive about the sheer steepness of the hills. 'We're not young,' he admitted gloomily.

However, as soon as we had our first look around, we knew. This was the one. Spellbound, I stared at the landscape tumbling endlessly in front of me. The cottage was sitting on top of steep rolling hills, with gullies and ridges and fascinating dips and hollows. The unbroken panorama stretched to where low hills rose in the blue distance – the landscape so big and open and infinitely awesome that it still expands our souls.

We sold everything and bought Clifftop Cottage Farm with its forty acres of hills and thistles. To celebrate our beginning of something new, we popped a bottle of a vintage red, and as The Engineer raised his glass, I cleared my throat and wiped a few tears from my eyes. Retirement was suddenly absolutely perfect. No need for road trips. No need for cruise adventures or retirement hobbies. Somewhere deep inside I knew that our retirement was going to be fulfilling, exhilarating and a real buzz.

TWO

We Celebrate Retirement

Sometime later came The Engineer's formal retirement dinner. This rather grand function was held inside a magnificent great hall at the rear of a heritage-listed mill. There was a proper meet-and-greet from hosts in stylish suits and formal frocks, and everyone was there: the young engineering crowd, the traditional engineers and their partners, the impressive HR people – and me. At first everyone stood around sipping their wine and chatting politely. I didn't say much, as different groups of engineers chatted on about various projects: the usual stuff like operational readiness, HR personnel, and asset performance management. I listened without much enthusiasm; I sometimes feel that engineers should get out more.

The meal was superb. The champagne popped as we stood around long banquet-style tables laden with selections of delicious cuisine and I soon found I was enjoying myself. Everything was going swimmingly until, at the peak of the evening, brimming over with enthusiasm and encouraged by the easy-going chatter, I spilt the beans and babbled on about Clifftop Cottage Farm on forty acres. 'This is our retirement plan!' I announced proudly.

There was an awkward pause as everyone stared. Eyebrows were raised and it struck me right away that telling about our farm wasn't the most sophisticated topic of conversation to be having at a formal dinner – in the city. A couple of worthy young wives, with drinks in their hands, had the audacity to dissolve into fits of laughter, while I stayed quite still with a silly smile on my face. It was no good trying to justify why I loved the notion of being a farmer; that the idea of retiring onto a little farm was as fascinating as it seemed ridiculous. I clasped my hands nervously. 'So just when I thought I was a retired teacher; we bought a farm!' I found myself gabbling on. 'I needed something more than my life being nothing more than a series of digital communications. I was spending too many hours on-screen – my face lit up by electronic light.' There were a few seconds of silence as everyone, it seemed, turned to look at me with a mixture of surprise and sympathy. After another pause, one engineering manager nodded and said, 'Oh.' And that was all he said.

Well, that was awkward. But then Tony from project management and quality assurance said, 'I want a retirement like you're having.'

And it seemed that old retired Eric shared my enthusiasm. He leaned over towards me and beamed. 'I think this is the beginning of a beautiful retirement plan. There's no better place than a farm. I grew up on a cane farm. There is nothing glamorous and every day you never know what you're gonna get, I can definitely tell you that.'

'Okay, here's the deal,' Tony said. 'The rule of retirement is to get yourself a design expansion application and make your farm extraordinary. It's really simple.'

I grinned. Yep! Got it!

THREE

The Realistic Hobby Farmer

Grandad used to say, 'If you want the best, you have to be the best.' My dream was to have the best farm, the best animals and the best outcomes. Such an idealistic view for someone who knew next to nothing about farming animals and had no practical hands-on experience whatsoever.

One of my earliest childhood memories is of holidaying with my friend Nellie on her dad's dairy farm. We made out we were cowgirls and played a game chasing poddy calves and jumping on their skinny backs to ride them around the paddock. This pastime did cause some chaos in the cattle yards and it was only a matter of time before we were caught by her enraged dad. And it was when we heard him bellow over the fence that I realised that this hoot of a game was causing some distress to the calves. That was my undoing, and as we tried to run away, I fell over a log on the ground and knocked out two front teeth. Stretched out on the grass bleeding, I sobbed loudly. I was banished, never to return again.

After that regrettable experience, my interaction with farm animals was limited to the occasional peer over fences when my parents took 'us kids' to holiday in the beach house up north. At the back of the property were some fields where friendly horses and ponies grazed alongside a few cows. Not particularly interesting. And there was another time, as a teenager, when I was on a school camp: without breaking a sweat, I organised a promising group escape into a nearby cattle property. We all lay down in the long grass to see if the cows would get curious and circle in to snuffle us. They didn't. That was as near as I ever got to farm animals as a kid, but even those brief glimpses had left an indelible stamp on my mind.

Despite my inexperience and lack of knowledge, I was ready enough to learn as much as I could about farming animals. At first, finding helpful, simple and complete information anywhere was a challenge. The internet is always a help and I frequently found myself on Google, jumping from one article to the next, digging up a scrap of information here and another scrap there.

At first, nobody paid any attention to the new owners at Cliffside Cottage Farm. Then word got around and I suddenly found that now we were living on the farm, The Engineer and I would be invited to pop over for 'a cuppa'. Everyone, it seemed, was really welcoming and ready to help, and the kettle would soon be boiling, as neighbours and local farmers with considerable practical know-how kindly shared their own stories and experiences.

I was naturally a source of great interest. People were clearly curious as to why a retired teacher, and a woman to boot, would want to work hard on a farm and we encountered a few amused glances. There was some discreet nudging and whispering when people realised that we are really quite old. I couldn't blame them. We had turned up out of nowhere and had settled on Cliffside Cottage, the worst farm in the district. Our property had a troubling history, and it was local knowledge that previous owners had neglected their farm animals. One chatty neighbour offered to show us where some Droughtmaster cattle had been buried behind our smaller dam.

'Stupid townie buggers running cattle,' he said. 'They thought Droughtmasters didn't need water and so the poor beasts died.'

Well, I was appalled.

Real Farmers! Real Stories!

Our nearest town is the sort of place where everyone knows each other by name, and in the centre of town is the locals' favourite hangout: the Bull & Barrel Country Pub. There's a sleepy charm about this old-fashioned double-storey hotel, with its upstairs ballroom, bare wooden floor boards and shiny chandeliers. It's a place in which everyone feels at home; a meeting place for anyone to sit and talk.

And the Bull & Barrel Café, really in need of a lick of paint and repairs on the rusted tin roof, squats in between the pub and the produce store. On the pavement level there are still a few rusty rings attached to the structure for the purpose of tying up horses and dogs. There is even the remains of a mounting block, used by past generations of inebriated stockmen to climb onto their horses.

Up a few steps from the pavement, the café veranda is enclosed with wrought-iron railings and French doors lead out into a beer garden, fenced off with old railway sleepers. In the footy season, retired farmers can be seen leaning back in comfy chairs, entertaining old friends and mates for hours with their stories and rich memories. And it's here that these tough old blokes regularly shed beery tears and gnash their teeth when a favourite team loses convincingly.

HARRI-HENRY'S FARM

Most country folk seem to drop into the café or the pub at some stage for a good old chat: regulars like the people who work on nearby farms; passing travellers who popped in once on their way home and keep on coming back; and truckies, dropping in from the busy highway, after they've discovered the fish and chips are worth a detour. We've met many down-to-earth, hugely likeable, tough old farm blokes and their chatty no-nonsense partners, all good for a yarn about how they 'do stuff'. Over a few beers they open up and share jokes and hilarious anecdotes about their farming practices. I love listening to them talking, and in a short period of time we've been able to learn a great deal about growing healthy productive stock. They share all the little things and shortcuts they've learnt over the years from other blokes and other places; and different ideas about what we should, or should not, be doing as new farmers.

The town has its fair share of unforgettable characters. We had only just moved into the farmhouse when, on an impulse one evening, The Engineer decided we should pay the Bull & Barrel Country Pub a visit. It isn't the most elegant of surroundings, but I had a warm sense of belonging sliding into place as we trod along the creaky verandas and into the musty-smelling dining-room. Inside, local paintings, pottery and wildly colourful beads are draped around the room in a sort of untroubled lack of order. I looked around in surprise at the road signs nailed to the walls – the ones that tell us there will be kangaroos and wombats for the next five kilometres! My favourite is the one from Australia's longest straight road which runs for 88 kilometres on the Eyre Highway in Western Australia, specifically signposted to warn of the presence of camels, wombats and kangaroos.

We munched through scrumptious blackberry muffins and sipped piping hot cappuccinos, as we listened to the happy chatter of voices and bursts of laughter over the noisy clink of glasses. Then we met Dusty.

'Evening. Good night for it.' Dusty noisily cleared his throat and introduced himself, taking off his cap and shaking hands.

'Food's good. Better than a pavement pizza!' And he allowed a smile to spread slowly over his face.

Over by the bar, Steph laughed with her ever-widening smile. She looked pleased to meet us as she bustled through from the kitchen to write up the next day's lunchtime menu on the blackboard. She turned to me with a sparkle in her eyes. 'I'm being interrupted by a parrot down the hill.'

Steph is the hotel chef and also cooks fine food for the café. She grew up around shearing sheds and has had a fascinating life as a bush cook, responsible for feeding ringers in stock camps, and as a station cook on a pastoral station. 'Such a hard life it was too,' she told me one day. 'Baking was a constant task. Hey, I even got married in a woolshed.' Steph's outstanding meals at the Bull & Barrel Pub quite belie the modest description on the menus. She makes great coffee and cooks up handmade French pastries and cupcakes. And her crusty breads are quite legendary.

She has some funny stories to tell and I spent a lot of our time laughing as she told of the uproarious happenings she said had gone on at the pub over the years. Like the parties. Wild ones where there'd be people sleeping all over the verandas for the night. And the time someone snuck Colin, a baby crocodile, into the pub's tropical fish tank. Everyone fed it hot chips until it grew too fat for the tank and then, one day, Colin somehow just happened to disappear. No one seemed to know where he'd gone.

Or the time when Eddie was convinced he'd heard someone playing the piano one night. But Dusty said, 'We don't have a piano here. Stone the crows, Eddie … you were pissed.'

'No, I weren't!' Eddie was quite offended.

Eddie is a funny bloke. He used to drive B-double trucks out west and has a rich history based around the consumption of booze. Wearing the traditional akubra, his thatch of fading red hair thrust inside, he appreciates being referred to as Mr Style. No one knows how this name came to be. But he is, I learnt, the kind of guy who doesn't beat around the bush. If you want an honest opinion, he'll give it.

And he did. I had suspected nothing when he decided to wind me up. We'd begun the habit of dropping into the Bull & Barrel Pub on Friday nights to sit in a cosy corner and dine on generous servings of fish 'n' chips. One evening, before we knew what was happening,

Chapter Three: The Realistic Hobby Farmer

there was a cheerful greeting and Eddie roguishly ushered us to our table.

'What ya eatin' tonight?' he demanded, staring at the blackboard menu. 'Try the curry. It's bloody yummy.'

Then, with a loud belching chuckle, Eddie flamboyantly renamed The Engineer with the grand title of 'Mr Needs a GPS to Survive'. His voice was very deep and loud. And when he announced that I was to be 'Harri-Henry, our Country Chick', the regulars drinking at the bar burst out laughing.

But our real favourite character is Earthy Murphy. Earthy Murphy is a retired cattle farmer and training greyhounds is one of his passions. Stories are told about how he is a 'few roos loose in the top paddock'. He is known as the bush mouse because of his reputation for falling sound asleep anywhere out on his farm. 'His mouth wide open and snores so loud the wallabies come over to check him out,' said Dusty with a grin.

We met Earthy Murphy as we went into the pub one evening. He was leaning on the counter and he turned and greeted us enthusiastically. Then, with disconcerting frankness, he looked me up and down and in a booming voice announced that, in his experience, old women farmers were as rare as rocking horse shit!

I like him. He is always warm and welcoming, full of nonsense, and always seen in shorts and thongs, even in winter. He is down-to-earth and kind, with the knack of knowing how to think up ways to fix any farm problem. He's had a lot of practice. He looked at me one time with that thousand-yard stare of his and bellowed, 'Hey. Harri-Henry. Mate! I was a young shit farm-kid who grew up in the bush. I could shoot snakes, change tyres, drive trucks and ride any horse.' He paused, and his face broke into a wide grin. 'Yep. Fell arse over tit again and again.'

Another time we met George. George is full of mischief and loves to goof off. After being a drover for many years, he retired to concentrate on endurance riding, then took up work as a contract fencer. He can play the mouth organ rather well and can knock out a rollicking tune, and one day Dusty kept egging him on to play some real music.

HARRI-HENRY'S FARM

'Classical music. Like Bach or something,' Earthy Murphy suggested helpfully.

Unfortunately for the rest of us, it turned into something of a comedy of errors, as George launched into a medley of unrecognisable tunes and went on and on – much to the amusement of the other diners. We all whistled and hooted and stamped our feet in delight, until Dusty grabbed hold of the mouth organ and chucked it over to old Uncle Harold James, who chucked it on top of the books that no one seemed to read, in the shelves either side of the old fireplace. We sat there in tears of hysterical laughter and there was a spontaneous burst of applause. Earthy Murphy laughed so much that he dropped his nearly empty glass to the floor. But it was all too much for Uncle Harold James, sitting at the far end of one of the dining tables. He fell unceremoniously out of his chair and after a minute of unrestrained laughter staggered out into the beer garden.

Chapter Three: The Realistic Hobby Farmer

George looked around in some surprise. 'Sorry mate,' he said. Then noisily he cleared his throat and bellowed, 'Won't happen again.'

He stared around at us for a moment, his face scrunched up in disdain. Then casually he strolled over to the book shelves, retrieved his mouth organ and lobbed it into the tropical fish tank. We watched in total silence now as the fish, disturbed by the bubbles, went to investigate the new toy. Then George turned to me. A trace of a smile flickered across his face as he said, 'Yo, Harri-Henry. Welcome to the Bull & Barrel.'

It was quite something to pop into the café one day and have someone introduce Aunty Tup. She was leaning over the counter chatting to Scooter, her head of spikey red hair tilted to one side, and I heard her bellow, 'What a load of bloody rubbish.'

Aunty Tup manages the café on Mondays and Tuesdays, and the rest of the week works part-time at the newsagent in town. She is middle-aged and instantly likeable, with a huge and witty sense of humour. Pretty much what you see is what you get. She has been called Tup all her life and can't remember anyone ever calling her Tuppence.

From our first conversation we became great friends, and the most entertaining story she ever told me was about a green tree frog named Froggins.

'It's a pearler!' Aunty Tup said, with her irrepressible gurgle of mirth. 'And I'm laughing now just thinking about it.'

I was intrigued. And it was a story that did have me in fits of laughter as she described what happened when her son, young Jim, squashed his pet green frog, grandly called Froggins, in his bedroom door. I can still remember her broad grin that day as she told me the story of poor Froggins.

> 'One morning, I heard a horrified scream from the kid's bedroom. "Muuum! Come quick! Froggins is broken!"
>
> 'We all rushed to his bedroom. All I could see was this squashed frog on the mat making a few pathetic squeaks. Its eyes were bulging!
>
> 'My diagnosis? One frog. One broken hind leg. And I said, "Oh shit!"

'I wiped young Jim's face with a piece of paper towel while he whinged and started to bawl. I tut-tutted a bit and muttered something like, "How about we ask Dad to get the vet to have a look at Froggins and fix him up?"

'I raised my eyebrows and looked really meaningfully at Big Jim. And in my attempt to sound confident, I said, "Better get Froggins down to the surgery and get them to fix him up proper."

'It was my wink-wink nod-nod moment. They were magic words and the kid was relieved. He coughed, wiped his eyes and watched me as I carefully, almost lovingly, wrapped that green frog in a piece of paper towel. Its froggy mouth opened wide into a comical little face – a sort of wide froggy grin.

"Now. You know what to do," I said meaningfully to Big Jim as I handed Froggins over. "It's got a broken leg; I'll leave you to deal with it."

'Well, clearly there was a communication problem because with reluctance Big Jim shrugged his shoulders and took the frog to the vet! I didn't think he'd actually do it. Take the frog to the vet, I mean.

'Well, it was several hours later when the vet phoned to tell me our frog was ready to come home. Incredibly, they enjoyed sedating him and deciding on the quite absurd arrangement to pin his leg and put a little plaster bandage on it.

'I was as angry as a cut snake. I slammed down the receiver and glared at Big Jim. "What the #?%?? have you done? It wasn't supposed to be a real trip to the vet."

'All over his face Big Jim wore this stupid grin as he told me he'd paid the exorbitant vet bill before carrying the frog home in a bright pink shoebox lined with purple and blue tissue sheets, with Froggin's scrawny bandaged leg protruding at an odd angle from his slimy green body.

'Big Jim actually looked surprised before he threw back his head and laughed and laughed. He did acknowledge his error of judgement.

Chapter Three: The Realistic Hobby Farmer

'As for little Jim, he was wide-eyed and spellbound. He stared at his little mate and shouted, "Froggins you're back. Dad, is he better?"

'As far as I was concerned, he was so lucky to be better. But all Big Jim said was, "Sure is son!"'

FOUR

Farm Advice and New Best Friends

With no warning at all, I fell in love with sheep and soon we had five Suffolk ewes and a ram mowing our front lawn. But since a few are never enough, I couldn't resist introducing some Southdown sheep into our little flock. Southdowns are gentle, manageable, funny, and frequently produce twins and sometimes triplets. Any thoughts The Engineer might have had of keeping me to a quota soon disappeared. He thought about twenty sheep would be nice. Now, even though I'm not a whiz at maths, it has become clear to me that, at last count, our sixty-one ewes, Fred the Ram, some twin lambs and triplets, are a lot of sheep.

Admittedly, it took some time before I began to discover that lambing is a bigger deal than I'd ever appreciated. Especially when things go wrong. These are the days I feel uninspired, and I catch myself scratching my head and checking my sanity. What once seemed like a passion-filled calling can turn into a bit of a slog. Stoically braving the elements is part of the deal and one of the downsides is trudging through the muckiest fields on stormy nights, dressed in extra-daggy clothes, heavy boots that squelch and splosh, thinking, *OK, this is harder than I thought.*

Chapter Four: Farm Advice and New Best Friends

So, it was just as well that I began to find there is always someone around who knows lots about farming sheep, and meeting Uncle Harold James was a huge relief. He is often to be found seated on a low stool outside the café, playing a guitar and singing old-time songs, the kind of tunes that tell stories from way back. He had once been a musician in a rock-style country band with other would-be country musicians – farmers who found that they liked playing music together around campfires and at small rodeo gigs.

'We sounded pretty good,' he told me much later.

Uncle Harold James just knows things. He has long been a farmer and, as it turned out, sheep were his speciality. Such a cool old guy with scrubby stubble on his chin, always wearing the same flannelette shirt and fluffy socks tucked inside heavy boots, a Bronco's beanie* plonked on his head, and yellow hay-bale twine twisted into makeshift braces to hold up his denim jeans. He had come through some hard times in Western Australia, and his bent knees show he's spent a lifetime in the saddle.

HARRI-HENRY'S FARM

'Western Australia? It's all the same,' he said. 'You drive for a day and nothin' changes.'

When Steph first introduced us, Uncle Harold was seated on his favourite stool at the end of the bar, conversing on all aspects of the footy scene.

'Harri-Henry's running sheep at Clifftop Cottage Farm,' Steph said conversationally. 'Tell her about the sheep station you used to run out the back of Roma.'

Uncle Harold James pursed his thin lips and rubbed his chin thoughtfully as he pondered over old memories. 'Well,' he said seriously, 'I don't think I'd know where to start. So, you want to farm sheep?'

There was something in his face that said a woman, and not a young one, would never make a go of it. He was an old man, a real tough old character, and I think he was trying to be kind as he bellowed in my ear.

Chapter Four: Farm Advice and New Best Friends

'You want to make it as a farmer? You'll have to stick it out and learn the ropes. The secret of farming is to know how to do things yourself.' He gave me a sideways frown and glanced over to The Engineer. 'And you think you can build yards and fix pumps?'

The Engineer and I looked at each other. 'Pretty much,' we agreed.

'Be in love with the land and your animals. Don't be shy of hard work.' Uncle Harold wagged a finger and his bright blue eyes twinkled at me. 'Be prepared for anything. It's not difficult. Just a case of learning to expect the unexpected. You're not going to find a magic wand to wave and fix problems. You just gotta be smart, farm girl.'

About a month later I was introduced to Mario out in the café gardens. He was sitting in one of the brightly covered armchairs, catching up with some mates and fretting about politicians, and commenting on how social media was taking over and pestering his grandkids.

'Way back in my day,' he sighed reminiscently, 'kids played outside and did stuff.'

Mario grows chickpeas, lucerne, wheat and potatoes on a large scale using an extensive irrigation system with pop-up sprinklers, all set by GPS into precise lines. It doesn't matter where you stand, you always see a straight line of pop-ups. He is locally known as The PM and eventually I heard the story. Dusty assured me it was true and that the letters stand for Parrot Man. It seems Mario copped a hefty speeding ticket out on the freeway. His excuse? That his very expensive parrot had caught its beak in a rat trap, and he had to take it to the vet for surgery! And so the name, The PM, has stuck.

Not long after this I met Scooter. It was a Saturday morning at the pub's annual Beef Cook-Off competition, held on the big grassy oval behind the pub. There were people everywhere; it seemed as if the whole town had turned up, burbling with conversation and laughter. The competition was a fundraising event for a local community project, restoring a long-corrugated iron shed. The plan was to turn it into a community centre for various functions, like old-time dances, garden fetes and weddings.

'It's just a bloody hut!' Mario was not impressed.

HARRI-HENRY'S FARM

The Beef Cook-Off tables, set up at the back of the pub, weren't much more than some trestle tables covered in white butcher's paper for tablecloths. There was a selection of scrummy dishes, all homemade creations using fresh produce supplied by Mario, and beef products from local farms. I could only put up feeble resistance when I began on the pan samples of meat and bacon pie, followed by savoury scones and cheese selections. And I was struggling unsuccessfully to eat my way through the biggest hamburger when someone smacked me on the shoulder.

'Hah, Harri-Henry! So, I finally get to meet the city sheila from Clifftop Cottage Farm. I heard you were bored out of your tree in the office.'

Scooter was regarding me with a kind of amused interest as he leaned forward and shook my hand vigorously for a moment. With my mouth full of hamburger, I burst into a coughing fit which completely took my breath away, and it was a few minutes before I could croak out, 'Yeah, right.'

Chapter Four: Farm Advice and New Best Friends

I gave what I hoped was my most winning smile and his eyebrows shot up as he welcomed me with his ever-present wide grin. I instinctively liked him. He is lean, long-limbed and very tall, with a face that has been well sunburnt, and is, I've since learnt, an experienced kick-ass cattleman who can ride tough horses; muster cattle in mud, rain and freezing conditions; herd sheep with his team of Border Collies; hold a decent conversation; and bake the best apple pies ever.

Dusty told me how Scooter got his rather unusual name – a good name for a cowboy who had been one of the best bull riders in the state. He was well-known for his winning technique of scooting off the back of the beast in a spectacular way, despite menacing attempts by the bull to fling him forward into the dust.

Knowledgeable in the ways of animals and the land, Scooter knows I have never run a farm before, and loves to tell funny stories and joke about my sheep. Full of fun, he makes me feel good, and sometimes it is really hard to keep a straight face. I've tried to make a good impression, but there is no doubt that my inexperience means I am invariably good for a laugh.

One morning, not long after we first met, Scooter fronted up in his old battered ute with grippy fat tyres. His manner, as always, was optimistic, and I was full of enthusiasm as our conversation ranged over various issues: worming and when to vaccinate; how to work a prolapse; what to do with a bloated sheep; how to stop a haemorrhage; and most importantly, how I would cope if I needed to deliver stuck baby lambs.

'How long will it take me to learn all this stuff?' I asked.

'It takes as long as it takes, farm girl.'

And in that one simple conversation about sheep, he provided me with a list of dos and don'ts, and invaluable words of advice, until I felt that I'd learnt enough to understand many unexpected obstacles – and some notable exceptions. Which was just as well because I needed to be totally prepared for things that could go wrong. Smart management can prevent serious issues and illnesses happening because realistically, farming is so often a matter of balance between the affordable and the practical. Relying upon a vet to come and solve problems is not a good solution, and vets acknowledge that by the time they get on the farm and despite their intervention, more often than not, an animal will die anyway.

The Truth About Lambing Season

Baby lambs are the best thing about sheep farming. They are full of charm and cuteness, and love to run wildly across our gullies and chase each other through the yards.

Our lambing season starts in winter, when cold winds bluster up our slopes and, more often than not, shredded curtains of rain drip from heavy low clouds. It can be so cold outside and almost without thinking I put on layers of warm clothes as I dress: heavy jumpers, a beanie, lots of socks and woollen gloves. Our day starts early with a steaming mug of coffee and biscuits. We need to be out on the farm whatever the weather, and traipsing outside on freezing mornings I have been known to tuck a hot water bottle under my jumper to keep warm. I have noted on Facebook that electric gloves and socks are the in-thing to purchase for cold seasons, and I must admit I almost put an electric vest into the digital shopping cart. But this is Queensland and our cold winters are short-lived.

Most lamb deliveries are unexceptional and, generally, it's easy to spot the signs that a ewe is about to birth. Her teats look swollen and she'll usually go off on her own somewhere to have her baby. In no time at all, a lamb is born and it's a fascinating experience to see how the mother pushes her nose at the lamb and begins a thorough licking to remove the covering membrane.

Newborn lambs are born covered in a liquid (amniotic fluid) that works as a conduit to exchange heat and cools their body temperature quickly. The ewe will have a natural instinct to lick off this fluid and reduce the rate of heat loss and, by doing so, stimulate the lamb to stand up and nurse. Once the lamb is able to stand, it is a tiny creature on a mission, instinctively tottering towards the nearest teat. For a moment the lamb gives a few no-nonsense thumps with its tiny head, but soon the little tail begins to wiggle as the drinking process begins.

Chapter Four: Farm Advice and New Best Friends

The usual birthing progression for a lamb is the appearance of two front feet, a nose, a head and then the body sliding out. There are times when some lambs can be extra-large and, in simple situations, I take hold of the lamb's front feet and pull. It may take a few pulls – with a few moments between pulls – until the lamb slides out, still encased in its protective coating. In most cases the mother will lick the newborn's face to free the lamb to breathe. But sometimes I need to get involved and wipe mucous off the lamb's nose and face with a soft rag.

Cold winter winds can drastically increase the risk of hypothermia in newborn lambs. The combination of cold weather and a wet birth coat is a worry, as lambs can quickly chill and become distressed. In these conditions we've learnt to act quickly and carry the wet lambs inside, where we rub each one all over with a fluffy towel, while holding it in front of our warm reverse-cycle air conditioner. It's quite surprising how quickly the lambs stop shivering and their little woolly bodies begin to feel warm and dry. Then back they go to their anxious mums for a feed of nutritious colostrum. This first milk is full of nourishment, antibodies and energy, and it's critical that every lamb receives this as soon as possible.

Every now and then baby lambs are born at the bottom of our steep property, where they are vulnerable to eagles during the day and feral and wandering domestic dogs after dark. Moving sheep and newborn lambs up the hill to the safety of the yards and barns can present quite a challenge. Our fields, with stony, grassy banks, have enough steep sections to warrant a decent level of fitness for me and The Engineer. If the lambs are born at the bottom of the hill, we will have to struggle on foot to carry the lambs up to the sheep yards, encouraging the ewe to follow on behind. But sometimes, despite our best efforts, the ewe can become confused or baffled, or whatever, and just as we reach the sheep yard gates, she can have a 'freaking out' moment and determine her babies have been left far behind. Putting her head down, she will run all the way back to the bottom of the hill, bawling loudly – quite oblivious to the plaintive cries of her lambs and my despairing yells as I sink to the ground in a groaning heap.

I was beginning to realise that sheep farming was not for the faint-hearted after I was told some horror stories about problem multiple births. I couldn't believe how things can go horribly wrong if lambs are twisted or stillborn; or when twins and triplets get tangled up inside and jam themselves like a cork in a bottle, with heads and legs all trying to come out at the same time. The challenge is to sort out where one baby starts and another one ends.

Uncle Harold James said there was nothing to be troubled about and proceeded to tell me how to disentangle the legs of one lamb from another. 'Bring a head and two legs out together,' he said. 'But make sure they belong to the same lamb, in the case of twins or more. This is an unavoidable part of your life as a sheep farmer, so don't stuff around too much.' He grunted with authority ... but he wasn't finished yet. He went on and told me what to expect when it comes to ewes and difficult deliveries, where a lamb may present in a breech position, and birthing assistance may involve pushing the baby back in and manually turning the lamb around so it can be delivered.

I felt nervous at the prospect of having to turn a lamb around inside a distressed ewe. It seemed like kind of a big deal, I thought, as I took a few deep breaths and hoped fervently that any awful situations like this would *never* happen on our farm.

I kept hoping, and as time went by, I forgot about the real possibility that things could actually go wrong. But little did I know just how involved I was about to be.

I suspected nothing out of the ordinary one morning, when The Engineer strolled outside to see what kind of day it was, and came across a panic-stricken ewe with a lamb's swollen head protruding from her vagina. It was a grotesque sight and I blinked in disbelief when I saw it. The distraught sheep was staggering around the small field and smashing its unresponsive baby's head on the rocks. My horror was inexpressible. It was my first experience of such a thing. The baby was well and truly dead, but wedged in position with the lips of the ewe's vagina squeezed tightly around the head. The tongue, blue and swollen, was drooping from its mouth.

Chapter Four: Farm Advice and New Best Friends

This was bad, and I had NO idea what to do. I had to think of something fast. I rang Scooter but he didn't answer the phone, and I didn't bother to leave a message. So I tried Uncle Harold, who did answer straight away.

'Mate! What's up?' he asked, and I told him what had happened.

Given the issue of the dead lamb, it was hardly surprising that his next question was about the ewe. Was it possible to save her? I didn't know, but was prepared to do whatever I needed to. He instructed us to fill a bucket with soapy water and, using a running hose, work the soapy water with the water from the hose over the dead lamb until we were able to push it back inside the ewe. And then – pull it out. It was awful; it seemed ages before we were able to remove the baby – a huge ram lamb. In some relief, the poor sheep tottered off into the barn and we kept her penned up for a few days until she made a full recovery.

As Scooter said deadpan, when I told him our harrowing story, 'After you've been up enough dry gullies, you'll recognise that experience is everything, and you'll be rockin' and rollin'.'

And after this awful experience I did feel almost ready to face anything.

*The maroon and gold beanie worn by supporters of the Brisbane Broncos Rugby League Football Club Ltd, an Australian professional rugby league football club based in the city of Brisbane.

FIVE

Sheep Do Funny Unexpected Things

Muddled Moments!

Most of our ewes have their babies quietly, almost without us even noticing. But some do it with breathtaking acts of originality. Part of our daily routine is checking up on the sheep and their lambs out in the paddocks. One morning, I stopped to lean on the fence to admire our increasing flock of ewes and big fat lambs, bunting and sucking, and then stared in some bewilderment at two ewes giving birth side by side, having their lambs simultaneously.

It was all a bit confusing, but I did assume that the sheep would recognise their own lambs. However, a few moments later, it became obvious that as they licked each other's lamb, as well as their own, and lovingly nuzzled both lambs, the more confused and muddled both sheep and lambs became. I was muddled and confused. It just wasn't a normal day.

Because the ewes gave birth in the same little hollow on the side of a steep slope, the lambs managed to roll around and slide into each other, and now it was not easy to make out which lamb belonged to which sheep. There was no way I could tell them apart and I could feel my eyes narrowing as I wondered how this situation was going to resolve itself. There is a risk when ewes have their babies at the same time that the lambs can get so confused, they can sometimes be rejected by both mothers.

But all ended well: for the next couple of days I kept both sheep and lambs confined to a small yard and it didn't seem too long before they seemed to amicably sort out which lamb belonged to which ewe. As far as I know, they ended up with their own baby!

Miss Mollie's Adventure

I could never have imagined that what seemed to be just a fairly ordinary start to the morning would turn into such a spectacularly different day. It was the beginning of lambing season. I was enjoying a cup of tea and a quick read of the morning papers, my feet hanging over the deck railings, when I looked up and blinked in astonishment. I certainly hadn't expected to see two giant wedge-tailed eagles sweep over the house and circle down, lower and lower, to hover over our bottom paddock. Laid-back but extraordinary, with their big beaks and ready talons, they were looking for all the world like some seasoned travellers.

For a moment I hesitated, before rushing inside to grab the binoculars. Through the glasses I could see Miss Mollie had given birth to triplets in the bottom paddock – two black lambs and a white one – wobbling at some distance behind their exhausted mother. I stared in horrified disbelief as the eagles began to circle just above the lambs. Miss Mollie was in big trouble and it was all I could do to not freak out as I weighed up the options. I decided on the most practical, as I hollered and yelled to get The Engineer's full attention. Without missing a beat, we cranked into action zooming down the hill to rescue Miss Mollie and her lambs.

The Engineer, determined not to be beaten, hurtled ahead on the tractor like a racing king, his shouts thundering down the valley and his big shoulders hunched forward aggressively. Retirees are not sprinters, and I felt I just couldn't go fast enough. But even my pace was astonishing as I tore down the hill with trepidation, frantically trying to keep up. It was one of those memorable moments and we arrived on the scene not a minute too soon.

Our fears were justified because the eagles began to glide leisurely just over our heads, so low that I could have reached up and almost touched their massive wings. We kept them spellbound with our antics as we recklessly scooped up the lambs and bundled them into a cardboard box secured onto the back of the tractor. The Engineer waved and cheered, his big hat pushed back on his head in triumph, as he roared back up the hill with three little lambs huddled in the bottom of the cardboard box.

Now the show was over, the eagles swung around and gracefully soared up and away towards the ranges, while I literally threw my cap in the air and hustled the worried and confused Miss Mollie up towards the sheep sheds. I like to think that she was grateful for her dramatic rescue. I was SO excited and quite out of breath when we made it to the top. Miss Mollie settled into the barns with her babies all safe and secure and, to my delight, she produced enough milk to feed three lambs without any supplementary assistance. A perfect ending to a marvellous adventure.

The Baby Fight

'And there was Harri-Henry going backwards so slowly I could have hit her with a dish of peas! And I shouted out, "Just keep going, you'll make it one day".' Scooter's face broke into a wide grin and his words were almost drowned out by a burst of laughter from The Engineer and Dusty. I knew it must have seemed wacky so I began to explain the rather bizarre circumstances that had me creeping up the hill backwards, one step at a time.

Just when I thought I couldn't possibly be surprised by anything sheep could think up, I'd come across two ewes having a humdinger of a fight over a newborn lamb. The rest of the mob had moved on and the dispute was in full swing. At first, I didn't know which ewe was the real mother and I had to closely examine the tussling ewes before I could identify the one with a blood-streaked bum.

This was not turning out the way it was supposed to for a baby lamb. I bent down and picked it up and the two ewes stared at me for a few seconds, breathing deeply. Then together, they bustled into the lamb in my arms, nuzzling it, licking it and pushing it around: I thought for a moment that I was going to land sprawled face down in the dirt. The only way I could protect the lamb and get it up to the yards safely was to do it backwards. And I did. Holding the lamb close to my chest, the two ewes bawled in my face as they followed me through the gate, over the hill and up to the yards. I was glad enough to reach the top but, by this time, I'd realised it was going to take more than gentle persuasion to stop the 'understudy mum' from pushing into the yards to pester the real mum and her lamb. It certainly took some effort for me to block her off and prevent her from charging through the yard gates.

I had a sad feeling that the second ewe, so desperate to steal a baby lamb, was probably the mother of a dead baby we'd found abandoned a few days previously. It had been one of those mornings: I had woken up to find that it was decidedly chilly and we had discovered the little dead body lying out on the cold ground. We think the frosty weather

Chapter Five: Sheep Do Funny Unexpected Things

may have been too severe for the newborn lamb to survive. This was the first time I'd experienced a sheep with any sort of grief issues and my theory was that she needed to steal another lamb as a substitute for her dead lamb. It was such a shame to see her looking so woebegone as she peered over the fence to stare at the new mother and her contented baby in the yard, and sadly, it was quite a few days before the grieving sheep gave up and re-joined the flock. This was one time I would have preferred twins as then there would have been two lambs for two sheep. Perhaps!

SIX

The Drought Effect

Drought is a part of farming life in Queensland. One year we experienced a particularly harsh period: week after week of clear blue skies with no hint of rain; a shimmer of heat haze; and hot sunshine cooking bone-dry paddocks. The change in the landscape couldn't have been more tragic, as lush green paddocks withered away into dirt fields, scattered with tufts of brittle brown grass. And when the breezes blew, powdery dust hovered over everything. It was a really tough time and brought about some of the most intimidating challenges.

And so, I always looked forward to Stumpy dropping in unannounced for a quick cuppa on the back deck. We would talk about many things, especially management and health of livestock in drought conditions. Stumpy used to run a sheep station out west but now spends his days in town, working part-time in the produce store and pottering around his small acreage property, mending fences and *doin' whatever else needs doin'*.

'Shit yeah, mate,' he roars with laughter. 'I'm as useful as tits on a bull.'

He always has useful things to share, and ideas on how to properly farm sheep. Or goats and cows for that matter. I never get tired of listening to him as he puts my concerns and theories about animal care into simple words and explanations, which even I can understand.

'Yep. When it's a drought, you gotta do what you gotta do,' he commented one morning. 'It's all pretty simple. At first, you'll try to do what's right, and screw it up. It'll hurt like a splinter in a sore thumb. Then you'll get in, get your hands dirty, break a sweat, finally get it right, and call yourself a farmer!'

'Bullshit!' I said. And we both laughed.

One day outside the pub, we bumped into Stumpy riding his horse along the footpath. He had been part of a crew moving a thousand head of cattle from Moree in New South Wales, to Mungallala in Queensland. For some time, the herd had been living on the narrow strip of grass between farm fences and roads, known as the long paddock, after they were moved off a drought-ravaged farm at Tumut, when there was no longer enough grass to graze.

'Yeah mate! It was hotter than a billy goat's arse in a pepper patch!' Stumpy turned and winked at me.

Mis-Mothering Issues

Meeting ewes' nutritional requirements in a drought and on a budget is a challenge. Our paddock grass is good and retains some value for quite some time, however, when there is no good green pick to be had, it becomes time to hand-feed. Hand-feeding is enough to keep ewes in good condition. But ewes carrying twin or multiple lambs are more vulnerable, as their demand for nutrition and energy to support multiple foetuses is high. And if tiny lambs don't get enough colostrum – the first thicker, yellowish milk from their mothers – in their first few hours of life, their survival chances take a nosedive.

Meggs, our local vet, warned that we might have extra problems with lambing in harsh drought conditions. Her practice is a rural one, used to the problems of farmers, and she is always on hand to share practical farming advice. Although I took note of what she said, I put it to the back of my mind when our first lot of lambs arrived. They were gorgeous – some single lambs and lots of twins. All was going well despite the drought, but not long after this, things got bad. Seriously bad. What bothered me and what I hadn't expected, was the number of cases of mis-mothering we were experiencing. I simply had no idea that every so often a sheep will display utter indifference, or worse, total rejection, towards her new baby. The simple explanation is that the ewe is obviously in trouble – perhaps unwell or just exhausted from giving birth – or she has insufficient milk to feed a baby.

Things went from bad to worse. Hardly a day went past without a baby lamb being abandoned, or lying dead in the paddock. It was all getting very, very worrying.

Uncle Harold James expressed his opinion succinctly. 'You're going to have to cop it fair on the chin, mate. In severe drought, the ewes don't want to feed their lambs. And newborns either just don't, or can't, suckle.'

One awful morning, I came across dead triplets. Two of them had been licked free from their birth sac, but the third lamb was still encased. There was no sign of the ewe which had abandoned them, and I couldn't help feeling that if I'd been present when they were born, I could have saved them. They all looked a good size and in good condition, but needed the sheep to finish the birthing process.

The early bond between ewe and lamb is crucial for lamb survival. Ideally, a sheep will spend some time with her new baby, licking it and helping the lamb to nurse, while building up a maternal bond. If this bonding process is disrupted, mis-mothering can occur, leaving the lamb vulnerable to neglect and starvation. When we begin hand-feeding there is usually a mad dash by the sheep to the hay line, and some ewes race off, leaving their vulnerable lambs behind. And sadly, sometimes they never return to look after them.

On Google I read about the Lamb Mortality Project undertaken by the Commonwealth Scientific & Industrial Research Organisation (CSIRO) in Armidale, led by a Dr Sabine Schmoelzl.* It was a four-year Australian project concerned with understanding why lambs don't survive, or why lambs are abandoned by ewes. The dead lambs in their research flock were analysed in an autopsy and, from the data gathered, it was quite clear that there were signs of a birth-related injury in at least half of all lambs that died. The study also showed that lambs which were not behaving properly, were a bit slow, or couldn't bleat properly, were often abandoned because the mum couldn't form the necessary ewe/lamb bond. Also, if there was a difficult labour, the ewe may have had trauma or been exhausted and unable to immediately bond with her lamb. This article certainly grabbed my interest, as one lambing season we ended up with eight poddy-lambs to hand-feed.

Tales from the Nursery

Lambs become poddy-lambs if their mother abandons them or, unhappily, they lose their mother because of illness or accident. The difficulty is deciding how to feed the new poddy-lambs. If the lamb has been abandoned, I can sometimes milk the mother so I can get hold of some colostrum for her newborn. The challenge is how to best nurture the lamb, with the hope that it will naturally want to suckle from its own mother. This may involve milking the usually unco-operative ewe, with the hope that the process will encourage her body to continue producing milk, which I can then bottle-feed to the lamb. At such times we might need to hang on for several weeks – or it might only need to be for a few days, while the ewe recovers and gains her strength and mothering instinct.

Every so often, after a few days and several attempts, we have been lucky and the lamb has been able to suckle from its mother. But more often than not, in our experience, we end up bottle-feeding the baby lamb.

If the mother has died then bottle-feeding becomes the thing. I do try to milk other nursing ewes, in the hope that I can get enough to feed an extra lamb, but usually there is not enough milk left over after they have fed their own lambs. If I have no access to real colostrum, I will purchase lamb milk powder available through vet supplies and produce stores.

Hand-reared lambs turn into the cutest pet lambs that follow me around, seeking love and demanding attention. Our daily routine involves things like early rising, washing bottles, preparing bottles of milk, sometimes milking an unco-operative ewe, and feeding hungry lambs at least five times a day. For a single lamb, or even twins, I use a strong rubber teat attached to a plastic drink bottle. At first the lamb tends to reject the bottle: it isn't the same as the real teat, and the milk, made up from a milk powder product, is certainly different from mother's milk. But after the initial introduction there comes an all-too familiar moment – a little tugging through the teat as the tiny tail begins to wiggle.

Ruby

One of the wretched aspects of raising sheep is what to do with abandoned lambs. When I first set eyes on Ruby, this sad little lamb was trotting around the pregnant ewes searching for a drink. I was trying to figure out what to do with her when The Engineer wandered into view.

'What am I going to do with this lamb?' I called out to him.

'Hang on a moment,' he said. 'This might work.' And he pointed to the far end pen where an older ewe was lovingly licking her newborn twins. I looked at the sheep and her new lambs for a few seconds.

'Hmmm. Do you reckon she'd take on a third lamb?'

The Engineer glanced down at the newborn lambs wobbling around in the hay and nodded thoughtfully. So, I took this as a yes. The newborn twins were still wet with birth coating and it took me some time to carefully rub the sticky substance all over Ruby until she was as wet as the twins. Then I stood back and watched as the little lamb began to huddle in close to the ewe. The sheep pushed her nose interestedly at Ruby and began to sniff her all over. And Ruby began to push into her udder with a singleness of purpose. I became cautiously optimistic and all afternoon I kept popping in and out to see what was happening. Just on dark I poked my head into the pen and shone a torch on the ewe and her *three* babies. The sheep and her extended family had settled comfortably into the straw and they all looked quiet and comfortable. I was beginning to feel confident, despite some tiny lingering doubts still messing with my mind.

Next morning, I got up even earlier than usual, almost at first light. I could hardly wait to see if Ruby was okay. I peered nervously over the gate into the barn, and I could tell straight away there was something wrong. At one end of the enclosure, the ewe and her plump little twins looked the perfect family, but at the back of the pen I could see Ruby's cold body stretched out on the ground. I felt so sad and, in my heart, there was a wistful regret that it hadn't worked out.

It had seemed such a good idea at the time and I had hoped that Ruby may have been able to grow up with a nice foster family. I made a mental note that I wouldn't make this mistake again.

Boris

As bad as this experience had been, there was worse to follow. The following day this miserable state of affairs happened again and I couldn't believe I was seeing yet another abandoned lamb – a determined knock-kneed little fellow bleating urgently for his mother – standing all alone in the midst of bossy ewes and lambs. The most heartbreaking moment was realising Boris was a twin. The ewe had abandoned both lambs and, to my disgust, the crows had already pecked the eyes out of the other lamb's tiny lifeless body. Nature is so cruel sometimes.

One of the difficult things with orphan or abandoned lambs is knowing just how far to go in order to look after them. Boris was a picture of misery and I knew that unless I took charge of him, he would die. So, Boris would be bottle-fed; it was the right thing to do and I wasn't going to mess this one up.

Admittedly, I shouldn't have gone to the pub and told Clay. Clay is a tall, muscular dairy farmer in his fifties and one of the most respected blokes in our district. He is often to be seen in the pub, chatting and laughing uproariously with his mates as they play darts in the evening.

Clay lifted his eyebrows when I told him, and the discouragement on his face was obvious. He leaned back in his chair, shrugged his shoulders, and admonished me sternly.

'Pet lambs! Hell no, Harri-Henry. They cost a fortune to rear on milk formulas. They never grow properly and you'll end up with skinny, scrawny wailing sheep – no good for anything. And *never ever* give them a name!'

'Of course,' I said hastily. I didn't know how to answer. Clay has become a good friend and I have always been appreciative of his efforts to make a farmer out of me. But I couldn't very well abandon this poddy-lamb – now named Boris!

Boris was easy to feed. A bottle of warm milk was a delight for him, and at the start, I was dishing out seven little bottles a day. Outside, the temperatures were hovering around zero and Boris looked so tiny as he curled up on his soft bed inside a dog crate I'd plonked down in the hallway of our warm farmhouse. In days gone by, when newborn lambs were found suffering from cold and exposure, the farmers used to wrap them in a dry towel and put them in the warming oven of the big old range stoves and fireside ovens, with the door open, where it was warm and cosy.

It wasn't too long before Boris needed company – a family! During the day he was lonely, and reluctantly I made the decision to put him outside with the older, bigger lambs and their mothers. But Boris was too small. He shivered in the cold and was distressed when the bigger lambs pushed him around and knocked him over in the dirt.

Chapter Six: The Drought Effect

Now I had two problems. Firstly, Boris needed a playmate. And secondly, he needed to stay warm during the day without a mother to shield him from the icy winds. I began with making Boris a coat. Two warm coats in fact, from a pair of my fluffy tracksuit pants, and he became the coolest little dude on the farm. However, finding Boris some friends was proving difficult.

One morning, he wandered away from the bigger lambs, found a gap in the wire fence and went roaming. He pottered through one of our more open dog yards where Hannah, our Belgian Malinois, was lazing in the sunshine. Without any warning, thwack, Hannah grabbed him! It was horrible; poor little fellow. She bit his face hard and he was bleeding. I felt bad as his ability to suck on the bottle was now severely compromised.

Yet most surprisingly, Boris seemed untroubled by this experience and it was rewarding to see how he began to put on weight, and his scrawny legs grew long and strong. I only had to call his name and Boris would do a mad dash across the little grassy yard and furiously bunt at my shoes, my legs, any part of my body, in order to get his bottle. I felt a certain smug satisfaction as I watched his first attempts to eat grass. My ongoing problem, however, was what to do with him. Boris still couldn't be housed with the other lambs and he was lonely on his own.

How Boris Gets a Family

It's funny how things work out. A few days later everything changed when a gentle old ewe popped out twin girls. I had a good feeling that this experienced sheep could look after her lambs. But it was not to be. Just when things seemed like they were getting better, they didn't. Any hopes I had of her managing to feed twins were soon crushed when much later I discovered that the twins had become triplets. Two girls and a boy. I was delighted. Even at this stage I was still hopeful that this experienced old ewe would manage to care for three babies. I left her licking and nuzzling and doing her thing. Later in the day, I again dropped by to see how it was all going and I was flabbergasted to find yet another lamb – number four. Another boy wobbling beside the other three tiny ones. This poor sheep now had quads to look after.

Quite enthralled, I sat down on a hay bale to think. I hadn't expected this, and I still couldn't believe it as I hugged four exceptionally tiny lambs. Then my elation turned to apprehension. It was plain that it would be impossible for this old sheep to cope with feeding four hungry lambs. She was obviously bewildered by all the high-pitched cries as the four lambs nuzzled into her woolly flank, and standing quite motionless, she fixed me with a troubled stare which said quite plainly, 'This is not working out well for me!!'

It was no good bringing to mind Clay's words as I considered having five lambs sleeping inside at night. Lots of lambs. Lots of bottles of prepared lamb formulas … and lots of work.

'What the hell are you doing now?' I could imagine Clay saying.

Still with quite some astonishment, I carried all four lambs inside our warm house to settle inside another dog crate next to Boris. And Boris was soooo excited.

During the day, all the lambs pottered around the backyard and at dusk, when the temperatures dropped and we all began to shiver, five friendly lambs filed into their large dog crate houses.

Scooter chuckled and shook his head when I told him about the quads – now to be named the Muppets. 'On ya, Harri-Henry. Just another day on the farm.'

I must have been looking a bit shattered because he added, 'You're not the only one bottle-feeding poddy-lambs. Old Merv down the road is feeding seventeen!'

Rosie and Daphne

The next morning another ewe dropped her baby and just wandered off, leaving her lamb, Rosie, toppling onto her face as she tried to struggle to her feet. As I gathered her up, I turned around and caught sight of another sheep with a newborn lamb and from what I could see, it seemed unusually tiny. This should have raised alarm bells, for later that morning I found yet another little lamb tottering down the fence line, trying to reach the sheep grazing in another paddock. A sense of hopelessness overwhelmed me when I realised that the second sheep had given birth to twins, then disowned one of them. It wasn't looking good.

I did still hope for the best as I put both ewes and three lambs into one of the smaller pens. I checked their teats to see if they had sufficient milk to feed their babies, and it was with some alarm I realised that their teats were limp and not plump with milk. My shoulders sagged lower. This scene was becoming all too familiar and was worse than I had feared.

A great weariness began to creep over me and, as soon as I could, I phoned Uncle Harold James. He listened to my tale of woes and said, 'Okay. It's unfortunately common in a really bad drought for ewes to reject at least one of their lambs if they are multiple bearing. Or sometimes twin lambs stray away from their mother in different directions. She may try to keep them together – or sometimes choose to only mother one, leaving the other to fend for itself. You'll find if she hasn't got big bags then there won't be enough milk for twins.'

'Is this because of the drought?'

'You've got no grass, and mis-mothering of lambs increases significantly when the ewes are being hand-fed. So, there'll be reduced milk production – and at the same time the colostrum quality may not be good. For the most part it will happen in the last month of pregnancy.'

I thanked him briefly and went back to the barn. The three lambs were miserable and hungry. And there was the matter of the twin that had been separated from her mother, the second sheep, missing out on crucial bonding time. Seeing that both ewes had no milk in their bags, it was clear the lambs would have to be bottle-fed. I leaned back against a hay bale to consider my long-term options. I could bottle-feed them and leave them out in the barn with their mothers, and I could pop in and out at frequent intervals with the required bottles of milk formula. Even in the cold evenings. Or I could take all three lambs inside the farm house to join the poddy group.

I put off decision-making until after I had successfully bottle-fed the lambs at both the late afternoon and evening feeds. I did think the mother of the twins was playing favourites, preferring her first lamb, Daphne, over the other twin girl. But in the jostling around of sheep and crying lambs, I dismissed it as my imagination. In the end, it seemed easier to leave the lambs with their mothers for the night. I told myself I could review the state of affairs the next day.

It was still dark the following morning when I got up before the birds to prepare the first round of bottles, then went to check on the lambs. I was just inside the barn door when I saw Daphne's twin sister lying dead on the ground. She was squashed up against the gate and I supposed that her mother had aggressively bunted her away. And to make matters worse, the other two lambs were lying huddled together in the straw, trying to get warm. The ewes were not interested in nurturing their babies and the weather was awful, with foreboding black clouds and drizzle in the air. It was too cold and miserable to leave Daphne and the other lamb, Rosie, in the barn, so I bundled them inside to join the poddy group. I realised I was taking on a long and time-consuming burden, but being mama to tiny cute lambs was becoming quite an obsession.

Buttons

Every now and then a lamb is born that is just so cute – and we got Buttons. She was gorgeous, with a big black nose and huge black eyes, like monstrous buttons, sewn into her white woolly coat. Buttons was one of the lucky ones; she had been mothered well and was off to a good start, and it was a pleasure to watch her springing and bouncing around in our big paddocks with the other lambs.

One morning, however, I got a nasty surprise. Buttons was lying all alone in the middle of the paddock. I had an uncomfortable feeling that something wasn't quite right as I walked over to her and got close enough to reach out and stand her up. Her big black eyes peered up at me as she sank back down to the ground.

'Bugger!' I groaned. This didn't look good.

The Engineer carried Buttons up to the poddy yards, where Boris, Rosie, Daphne and the four Muppets were lined up at the gate staring with great interest as we made a special place in one corner of the barn for the newcomer.

Lying down beside a bowl of water and some lucerne, Buttons stared up at us, her eyes bright and alert. We considered some possibilities for her condition. Did she have a tick? Maybe a snakebite? A disease? A traumatic injury?

We found no signs of injury, no evidence of a tick, and a snake bite seemed unlikely, as she was still very much alive. I was baffled. She hadn't lost her appetite. However, being unable to stand meant she couldn't graze, though I was relieved to see her nibbling at the lucerne.

The following day she was worse. I tried heaving her up to encourage her to stand, but almost immediately she would sway and wobble and slump back down again.

Chapter Six: The Drought Effect

'So, what's the plan?' The Engineer enquired.

My plan? I had no plan. I was still bottle-feeding the poddy-lambs, so the thought occurred to me that perhaps a plan might be to bottle-feed Buttons as well. This was about the best solution I could think of. I sat on the nearest hay bale, plopped this big lamb across my knees and began to force-feed her a bottle of milk. Her black eyes stared as she wriggled and objected, and it seemed to take forever for the bottle to be emptied.

Each day I continued to force-feed her three big bottles to cover breakfast, lunch and tea! Her condition remained a mystery, though I did discover large lumps either side of her neck. Could these be swollen glands? It was all very bewildering.

At first, I was horrified at the prospect that she would never be able to walk again, and day after day, after the ordeal of feeding her a bottle of good milk, I would massage her legs and stand her up. But she continued to flop to the ground, unable to get up without assistance. And every time I heaved her up and held her, she would pee then collapse, sagging back down to splay out on the dirt! Over the next few weeks, she remained the same and my hopes for her recovery grew rather dim. What was I hoping for? That after time she might just recover on her own?

Despite her deteriorating condition, I decided to keep going the way we'd begun. As I bottle-fed her, I cheerfully sang songs – kids' nursery rhymes like 'Harri Had A Little Lamb' – as we struggled through our daily bottle routine.

Things did not improve and I had my doubts as to whether she would ever recover. I was never going to tell Clay about this lamb! But it was all too much for The Engineer; he told Clay.

'Now what's she doing?' Clay asked with evident disapproval. 'Once an animal becomes immobile it's pretty much doomed.'

These grim words from Clay depressed me even more. 'That's nice,' I grumbled.

'Clay's right. As long as a sick sheep stays on its feet there's a chance of recovery. Being upright keeps everything working.' The

Engineer spoke gravely and his good-natured face was looking concerned. I slumped into an armchair. At least I didn't argue: I just wasn't ready to listen.

Buttons stayed this big floppy lamb for what felt like forever, yet I forced myself to be optimistic. The naughty poddy-lambs, all seven of them, took enormous liberties with Buttons while she didn't have the strength to protect herself. At times she could hardly lift her head as they jumped on her and over her and tried to push her out of the way. Fortunately, her big size was a match for these tiny playmates.

Probably a month later, something unexpected happened. I began to notice that each morning when I reached the barn, Buttons would be sprawled in a different corner of the pen. My guess was that she was attempting to follow after the other lambs. She seemed bright enough as she greeted me with her head up, but legs folded beneath her. So, curious to know more, I began putting her in the furthest corners of the barn each evening and noted where I found her positioned the next morning. There was no doubt about it. Somehow, she was manoeuvring herself from one end of the pen to the other, in order to be close to the poddy-lambs. My hopes grew. There was no further doubt. This big woolly lamb was getting better.

Then not long after this, I arrived at the barn one morning to see Buttons stand up and wobble a few steps. I could hardly believe it! It was immensely satisfying, and over the next few days I saw some slow and steady progress. Buttons began to stand on her own and stagger around behind the other lambs, munching into hay and sheep nuts. Whatever the cause of her mystery condition, the undeniable fact remained – she was up and walking, almost trotting, after the poddy-lambs. She was going to be okay. For another week I persisted with her milk bottles until the final rejection came. She bit the top off the teat. I agreed, enough was enough.

Another month later, the difference was dramatic. She had put on weight and was rollicking around the yards with the poddy-lambs as if nothing had ever happened. It was incredible and I couldn't wait to tell Clay, my face beaming with a self-satisfied smirk. He didn't share my enthusiasm. He is annoyingly practical sometimes, but he did have a rueful half-smile on his face when he saw eight lambs bouncing across the small paddock. And he did acknowledge her recovery was all rather surprising … although he laughed at me with his booming laugh and said he couldn't call her Buttons.

Strolling through the yards with eight friendly, bouncy, woolly lambs at heel is a most satisfying experience, especially as I know that, without intervention, all of them would have died. This is good on so many levels: they are cute and remarkably healthy considering their early days on formula milk. They've bonded together well – and Boris got a family.

Postman's Hay

It seemed that it might never rain again, as the drought continued with far-reaching consequences.

'So, when's the rain coming?' It was all we ever talked about. All we ever thought about. It was all very worrying.

'Anyway … today is a day closer to the rain that's coming!' Scooter tried to put a positive spin on it.

It was hard enough paying over-the-top prices for feed, but then we began to hear worrying rumours. Bores were drying up; creeks were dry; and underground water of any quality was disturbingly low. Because of this state of affairs, crops were not planted and there was little chance of any winter feed being grown. This was the story everywhere and it wasn't long before we had to take it very seriously indeed. Even the local Veterinary Teaching University was having emergency crisis meetings to work out where to source feed for the many animal groups on their farms.

Our challenge was finding enough feed of any kind to maintain our assortment of breeding ewes, rams, eight cows and six goats, and I joined others in panic buying whatever we could find. I knew I was on borrowed time and the day came when our only option was to reduce our stock numbers. It was tough and I tried to look okay. But it took a definite effort of will for me to choose which breeding stock to keep and which animals to sell. How many sheep could we realistically carry through into the winter, when grass grows slowly? Our sheep mean a lot to me and I knew most of the older ewes, having grown them from lambs.

So many farms in our district were having to reduce their stock numbers and sell off the animals they couldn't feed. Sale yard prices for cattle and sheep were shockingly deflated and we ended up selling some fat lambs and top breeding ewes for a third of their real value. But there was no time to dwell on this as I still needed to figure out how I was going to feed our remaining sheep and goats, three cows and Christmas, the calf.

I began to explore every avenue. I was having great difficulty in finding enough bales of anything in our local area, and I was torn with conflicting emotions as I realised that things were getting pretty desperate.

One really hot morning, I wearily walked out to meet the postman. I was glad enough to see him at the gate with a parcel to be signed for, and I needed someone to whinge to.

'I'm having a bad day,' I grumbled.

'Oh yeah, what's that?' Derek glanced at me for a moment as I replied honestly that I was worrying about finding enough feed to get me through the winter. Or at least, until some decent rains came.

'How much hay do you want?' Derek cast a roguish look into my troubled face.

'How much can I have?' I tried to keep my voice casual but my eyes widened in astonishment. And a quivering at the knees was the only give-away of my pounding heart as we discussed quantities and price for some old, but still good, stored hay sitting on his farm inside a massive timber shed. I didn't know Derek lived on a farm. He looked just like a postman. What *should* a postman look like, I asked myself?

I felt like bursting into hysterical laughter and after Derek drove off, I rushed inside to tell The Engineer.

'You're just not going to believe it!' I said incredulously. 'I'm buying hay from our postman.'

He lifted his eyebrows at this. 'Really?' And a look of the blankest amazement came into his face. 'You're buying hay from the postman?'

And we did. I was so overwhelmed and couldn't wait for The Engineer to go and pick up several truckloads of hay and take it down to be stored in *our* hayshed. Having grassy hay had become a precious thing.

And Then There Were Five

There was one time I came across a sheep standing in the far corner of our barn with an exceptionally tiny new lamb at foot. She didn't seem to be straining or unduly distressed, and there seemed no obvious cause for alarm, but I couldn't get rid of a niggly feeling that something wasn't quite right. All the same, I sat down on a straw bale to keep an eye on her. I was enjoying the moment, taking some photos of the lamb, when sure enough my misgivings proved to be well-founded. I could clearly see one tiny hoof and a nose peeking out from the ewe's butt, and this was somewhat confronting. I had a moment of panic as I remembered my lambing chats with Scooter, and how he'd told me that two feet and a nose are needed for a safe, successful lamb delivery.

Here was another thing: there was nobody around to help, and I dared not think beyond the fact that I had no option but to attempt to push the lamb's head and hoof back inside the ewe. It took some real effort as I gritted my teeth and pushed steadily, leaning hard against the ewe, until the lamb was inside again. Then, I groped around desperately until I could feel two tiny feet and a head. The sheep moaned and pushed roughly against me and, finally, as she gave a huge heave, I was able to manoeuvre the lamb out and he slithered onto the ground. I felt his slippery body as I wiped his little nose so he could breathe. Then he twitched and opened his mouth, as he tried to stand up on wobbly legs.

I leaned back against the bale of straw; I was feeling quite stuffed, but relieved that the crisis was successfully over. I began to take more photos as this second tiny lamb began to totter towards his mother and tiny brother and, as I did so, I gave another quick glance towards the ewe's butt. And again, I simply couldn't believe my eyes. Another nose was coming into full view, thankfully with two feet this time. And a few minutes later, this weary ewe delivered yet another tiny lamb. At last it was all over and the tiny triplets began nuzzling into the ewe's teats. I felt great, as if I had done all the work myself, and I was quite ready to face anything after this unbelievable experience.

But the smile was soon wiped off my face when the sheep began to prevent the triplets from suckling her teats, despite their persistent attempts. I didn't know what to do. By now the lambs were hungry and it was becoming a point of great concern that these tiny babies needed to drink the necessary thick, rich colostrum without delay. So, armed with a jug, baby lamb teats and a plastic drink bottle, I gently milked the ewe and with a feeling of intense relief, I managed to get enough milk to bottle-feed her three hungry lambs.

Things took a severe turn for the worse over the next couple of days. The ewe went downhill fast and was distressed to the point of collapse. She made no attempt to move much, although she would sometimes lie down beside her lambs to keep them warm and comforted. It was my responsibility to milk her and bottle-feed the triplets, but this sad solution lacked practicality.

After a few days it was clear that there was nothing I could do to help her, and with a feeling of inevitability, I reached the sad decision to ask Scooter to shoot her. It wasn't financially viable to involve our vet and when we discovered two dead lambs still inside her, it became clear that the simple explanation for her mis-mothering behaviour was the fact that she had been carrying five babies. Our triplets were actually quins. And, because the two babies inside the sheep were well and truly dead, and had been so for some time, there was no chance that a vet would have been able to change the outcome.

It was soon plain the worst was over and the triplets began to thrive as I fed them a commercial lamb milk powder product. The lambs loved me; they loved their bottles. In the space of a few weeks they grew bigger and stronger, and it seemed a good idea to introduce them to a lamb multi-feeder, because two bottles in two hands for three over-excited ram lambs is an almost impossible feat. So, a multi-feeder became an essential item in the lamb pens. It didn't take long before all three lambs had the new feeding process well and truly under control.

I thought all was going famously, but it was too good to last. It was then that I was confronted for the first time with the issue of bloat in hand-reared lambs. The little fellas were growing well and then suddenly, it was *Oh shit!*

One of them became ill, bloated and died. It was so quick, it didn't seem at all possible. It was the most heartbreaking moment and I was devastated. What should have been five babies had been whittled down to only two – now nicknamed This 'n That.

Ref: www.people.csiro.au/S/S/Sabine-Schmoelzl

SEVEN

Getting into the Nitty Gritty of Animal Management

Bloat in Hand-Reared Lambs

'Ya mustn't overfeed poddy-lambs.' Uncle Harold James came straight to the point. 'It'll kill 'em quickly!'

His words struck home as he told me that nursing lambs seldom bloat, however, bloat is a relatively common health issue in young bottle-fed lambs. The actual cause of bloat is an overgrowth of gas-producing bacteria, which can happen when lambs drink too much too quickly, and this can result in heart or lung failure, causing the lamb to die. At first the affected lamb will seem dull and not want to drink, and typically present with a swollen belly. By this point, alarm bells should be ringing.

I hadn't realised the significance of knowing *exactly* how much each lamb drinks, in order to prevent overeating, which can lead to bloat. Young bottle-fed lambs are not meant to have large feeds a few times a day – just a constant supply, little and often. Some lambs are greedy, and some hardly drink much at all. The drawback with *our* multi-feeder is that we cannot monitor how much each lamb is getting, as the teats are attached to a central chamber of milk.

I was really interested to view a video clip on Facebook, showing a long line-up of calves, all drinking from individual wine bottles filled with milk. This ingenious calf 'multi-feeder' had a wine bottle for each calf, and each bottle was individually attached to a long wooden plank. A controlled and well-thought-out, monitored process. I told myself that next time we had more than two lambs to bottle-feed, The Engineer could make us a wine bar/milk bar arrangement, just for lambs.

I'd also heard that feeding cool milk to bottle-fed orphan lambs is a good thing, because warm milk can be a culprit, providing bacteria with an ideal environment for fermentation.

Megg's response to that piece of information was clear: 'Well that's bullshit.'

On the other hand, everyone seemed to agree that it's critical to ensure the lambs don't drink too rapidly. And by using carefully selected hard teats with small holes, greedy lambs can be prevented from 'pigging' the milk. Old corroded teats, or teats that are overly soft, are a problem as they can allow the milk to flow too quickly.

And it seems that healthy lambs can be weaned off bottle-feeds from four to six weeks of age, providing they weigh in at about a recommended nine kilograms.

What Exactly is Bloat in *Adult* Animals?

Bloat in livestock is a common enough problem. The condition can kill ruminant animals such as cows, sheep and goats very quickly – often in a matter of hours. And the first indication of the problem may be a dead animal in the paddock.

Bloat is an excess of gases in the rumen, or first stomach, which makes the left side bulge outwards, and when the condition worsens, the bulge appears on the other side as well. Distressed breathing becomes an obvious symptom and the animal looks miserable and won't eat. If not assisted, there will be so much pressure on the lungs that the animal will die from suffocation.

The animals that tend to bloat are usually the greedy ones being fed grain or grazing in lush green paddocks, running from plant to plant to take just the best picks, instead of eating all the stalky underneath bits. And Scooter's best advice? When introducing livestock to grassy paddocks, grain or leafy lucerne – do it gradually!

The issue of bloat in adult animals had my full attention. I had not previously given this matter much thought, but then we had an experience with a goat with bloat and it was horrible. And, it's fair to say that by the time I was thinking that I needed to do something, it was too late for our goat.

To make matters worse, it seems that treatment outcomes for bloat are usually poor. On Google, I did find mention of old-timers using a bloat needle, or even a knife, to puncture the rumen and allow gas to escape. But this is considered to be a dangerous and drastic step, only to be taken if the animal's life is in immediate danger.

Veterinarian assistance is always the preferred option. A vet can administer antibiotics or attempt to deflate and de-rotate the stomach by piercing it with a needle under local anaesthetic. Alternatively, they may try to pass a stomach tube in to release some gas, then get some anti-bloat oil into the animal that way. But, we were fast learning that farming animals is so often a matter of balance between the affordable and the practical. That despite the intervention of a vet, more often than not, an animal will die anyway.

So, What Was My Paddock Bloat Plan?

The circumstance of our goat with bloat was disturbing and I needed to think about a future practical management plan. I was open to suggestions and I agreed with The Engineer that we should call into the Bull & Barrel Pub one evening, to have a discussion about bloat issues with some of our farmer friends.

It was nearly a week later when we strolled inside the busy pub. The dining lounge was full of people, and we could see Scooter having dinner with Dusty and Eddie.

'Hey there,' Scooter said as we strolled over and sat down. 'You finally figured out where to get a good steak?' He ordered coffee and

Chapter Seven: Getting into the Nitty Gritty of Animal Management

we chatted away, catching up on all the local gossip as well as grumbling about the hot weather; how the price for a bale of hay was going through the roof; and the ridiculously poor returns on cattle at the sale yards.

A good while later, I felt it was time to ask my question. 'Tell me again, what should I do if I come across a sheep with bloat?'

As might have been expected, this turned out to be a confronting conversation. My enquiry brought about a lively debate that almost got a little out of hand … and the general consensus made my eyes water. I could hardly believe what I was hearing. Seriously, it seems their quick cure for an animal with bloat is to take a carving knife, or similar tool, and stick it straight into the stomach – to pierce the significant bulge on the animal's left-hand side.

'Really?' I remained entirely unconvinced and couldn't stop giving them a look of reproach. It was no good attempting to raise objections. It has always been this way for the old farmers, and I get that. But I was unimpressed with their casual attitude.

'You reckon you could do it?' Eddie said in some amusement. And I had the silly feeling that they actually did think I could do it.

'You've got to be kidding!' I replied without hesitation. In spite of their assurance that this was a reasonably safe procedure, I could not imagine me plunging a knife, or any such thing, into one of my animals. However, the uncomfortable fact remained: I still had no plan if we experienced bloat issues.

I met up with Stumpy the next day and we talked a bit about cows and the drought, and farm issues in general. I murmured my thanks when he offered me a piece of his home-baked carrot cake. I always say yes to any sort of homemade cakes and I encouraged him to talk about how he'd learnt to cook. He told me how he had learnt as a kid to cook delicious apple pies and treacle puddings in his Grandma's bush oven. I was enjoying our conversation and the cake, but eventually I had to raise my concerns about paddock bloat, and Stumpy laughed. Folding his arms across his chest, he began to tell me stories about blokes who have saved cattle by shoving a sharp knife straight into the stomach.

'I really don't know, Stumpy,' I said a little doubtfully. 'I know these things are emergencies – but I'm not sure I could ever do this.'

'It's quite easy if you have a wide blade, mate.' Stumpy was quite carefree about this process.

Derek was also disarmingly candid. His postie van often chugged up the road past our gate, and next day when I leaned over the front fence and waved, he stopped to chat for a few moments. Now that I knew he was really an old farmer, and not just a postman, I most certainly felt I could ask what he thought I should do about an animal with bloat, as I was forced to confess my inability to cope with the schemes suggested so far.

'Now then, no buts,' Derek assured me. 'It's a good thing to have at least one practical person on the farm and you know common sense must always prevail. Get yourself a sharp knife. Or a tool like a piece of poly-pipe with roughly a 45-degree angle at one end. You can shove either the knife, or the tool, into the point of the bulging stomach and you'll hear a whoosh as the air rushes through. And stand well back or you'll get sprayed with muck!'

He rummaged through the glove compartment of his van and pulled out a heavy pocket knife, and I couldn't drag my eyes away from it as he told me how, when he was just a kid, his family moved out west to manage a cattle property. He can clearly remember the first time his dad fixed a bloated steer using a pocket knife and a spray can of disinfectant.

'He fixed it up pretty good. Dad did a lot of these paddock operations over the years and in most cases the cows were fine afterwards,' he said, a laugh in his voice.

After Derek had driven away, I spotted Ross riding his horse towards our farm. Ross is the town farrier and I had met him a few times through our friendship with old Uncle Harold James. This was too good an opportunity to pass up and, after we had exchanged a few niceties, I took a deep breath and screwing up my face, I asked him about bloat issues. And whether, in his opinion, I would have to shove a sharp tool into the animal's stomach to release the gas.

Chapter Seven: Getting into the Nitty Gritty of Animal Management

He gave me one of his looks and I had the feeling that, in his opinion, these things were quite common, and using a knife to plunge into a cow was nothing to be worried about. 'Stuff it, Harri-Henry. What have you got to lose?'

'I suppose if I must, I must,' I said, trying very hard not to show I felt impending doom at the prospect.

Earthy Murphy was the most direct. Leaning casually against a rusty cement mixer out in his yards, he chuckled as I looked at him questioningly. 'Harri-Henry, what the hell. It's life-giving to the beast, so don't make a big deal about it.' He nodded with good humour. 'You're going to have to suck it up and do it!'

In contrast, George was a man of few words. 'Shove a long syringe needle into the animal with bloat,' he said conversationally, 'not a knife. The needle only makes a tiny hole – although it takes a lot longer for the air to escape. But it's quite an easy thing to do.'

'Yeah, okay.' I gave him a wan smile as I tried to give the impression that I was almost convinced and that if I had no other course open to me, I might be up to it.

It was all very unsettling and I was still unhappy so I went off to ask Meggs for her veterinary opinion. I voiced my doubts only to find that Meggs was an enthusiastic advocate of the knife approach. I looked at her thoughtfully as she even went as far as showing me exactly where to plunge a knife, using her own obliging Border Collie as a make-pretend sheep. I was quite surprised. 'I never thought you'd say something like that!' I said.

When I got home, I sat for some time with my chin in my hand, pondering the matter. My first impulse was to side-step the issue altogether and stick grimly to my hope that this might never happen on our farm again. But my spirits sagged lower and this unfeeling response from Meggs sent me back to talk it over with The Engineer. 'Well, now you know what you're signing up for when you take on farming grazing animals,' he said unreassuringly.

Three weeks passed and I could not get around the fact that I was still stuffing around trying to figure out a bloat plan. Then one day Clay dropped by and I gave him a glum look as I told him my story about our goat with bloat, and the bottle-fed lamb, and my ongoing concerns with managing bloat, and that I still didn't have a bloat plan, and that there seemed to be no other course of action open to me but to …

'Hang on a minute, Harri-Henry, this is your lucky day,' Clay said soothingly. And with a real twinkle in his eye, he said all I had to do was to give him a call any time a sheep or cow showed any symptoms of bloat, and he would come and fix it for me. I believed him. Wow! How simple was this plan! I felt immensely grateful and I laughed and waved goodbye as Clay jumped on his horse, looked back with a grin and shouted, 'So long boss!' as he cantered gracefully down the driveway towards our front gate.

I sighed contentedly. I knew I could take Clay at his word. He's the kind of bloke who will step in when my courage fails me and by all accounts, it seems the animal will live, but will certainly die otherwise.

More Stuff to Know if you Love to Farm Animals!

Over time it seemed I was constantly absorbing a vast amount of animal husbandry information, which sometimes turned out to be downright nerve-racking. Farming is a skilful business and learning to spot trouble right at the start is of huge importance. I was aware that my inexperience would land me in trouble sooner or later and so I looked forward to our late-night conversations with Scooter. They are always enlightening: he can talk at length on most things to do with farming animals, and I'd mentioned one time that I'd bought quite a few bags of sheep nuts to help keep our pregnant ewes in good condition.

Scooter raised his eyebrows. 'You do know that sheep can fall over and not get up if they eat too much grain?'

I began to laugh. 'You are not serious!'

'Yes, I am. It's known as grass staggers.' Scooter spoke gravely.

'Oh?' I said rather surprised. 'So, what do you mean?'

'A sheep can fall down, go into convulsions or just die without warning. This can be grass staggers or milk fever.'

'What should I do then?' My face stared at him with blank confusion. Seriously, was he making this up? I honestly thought the story couldn't get worse. But it did. He began to explain in tedious detail about milk fever, something called grass tetany, and twin lamb disease. And I groaned. There seemed such a lot to learn.

Full of enthusiasm, Scooter chattered on at length. He dealt learnedly with the topic of hypocalcaemia.

'This happens when sheep have a calcium deficiency just after delivering a baby, and it causes collapse and a progressive coma,' he pointed out.

'What are the symptoms?'

'You might think the sheep is drunk. Her eyes will look heavy, maybe blinking, and she'll seem a bit dazed. Sometimes there's discharge from her nose and her breathing might sound different.'

'How different?'

'Sort of bubbling. Or makes a rattling noise!'

I must have looked doubtful. It was no good telling Scooter I was finding all this information difficult to process.

'Stop your worrying, Harri-Henry,' he said reassuringly. 'And get yourself some Metaject. The produce store will have it. It's ready to use and can fix up sheep within minutes. If you see a sheep down and she can't get up, and she's not in labour, whack in a needle and give her a shot. Just under the skin – on a leg works well. And the sheep will recover really quickly!'

After this conversation with Scooter, I Googled 'grass staggers' and found pages of involved information and lists of causes and fixes. And grass staggers, or grass tetany, led onto other issues for pregnant and nursing sheep, such as pregnancy toxaemia.

A symptom of grass staggers is the sheep staggering. It occurs when blood magnesium levels fall below a critical level, or when animals are running on pasture which has low available levels of magnesium – or as a result of increased body demands for magnesium during lactation or late pregnancy. The term 'hypomagnesaemia' is used for this condition.

Common causes are grazing animals on grass-dominant pasture or lush cereal crops, often without any hay supplementation; cold, wet, windy weather with little or no shelter, resulting in short periods of fasting; poor nutrition or fat sheep losing condition. Or sudden exertion and stress can bring it on.

A pregnant ewe needs her body's energy requirements met through good nutrition, especially in the final two months of pregnancy. But if feed quality is low, then the sheep can succumb to pregnancy toxaemia, which is known as twin lamb disease. Other causes of twin lamb disease include stressful events like worms, foot problems and mouth problems.

This is all seriously scary stuff. But there was no need to worry I decided, because I felt I was doing rather well. Our sheep seem to do okay and don't present any of these symptoms and this perhaps means that our sheep management in terms of nutrition, shelter needs and a stress-free life are providing a 'stagger free' environment.

On the other hand, if I come across a sheep that has fallen over, is unable to get up, has a drunken nodding of the head, and is dribbling, I will follow Scooter's practical advice and inject a shot of Metaject. We have some in the back of the fridge now. It comes as a complete package with instructions, a tube and appropriate needles.

As Uncle Harold James confirmed, 'There y'are, darl. The sheep will be up and back to normal within minutes.'

Lamb Pneumonia

I began to get better at recognising early stage symptoms of illness in sheep, and one day when I was leaning over the sheep yard rails, I spotted a young lamb just standing around, not attempting to graze or even suckle from its mother. It just didn't look right and I wasn't sure what I should be doing. The following morning, my imagination was giving rise to all kinds of anxieties and I was just about to phone our vet to ask questions, when the lamb died. The following afternoon, with some alarm, I found another lamb showing the same symptoms, and it also died in the night.

Chapter Seven: Getting into the Nitty Gritty of Animal Management

I sat at breakfast the next day lost in thought as I worried about the rest of our lambs. According to Google, it seems the two lambs that died may have had pneumonia. It was no good sitting at home moping about it all. I needed some real advice and so I drove around to visit another farming friend, Ted Graham.

'Ted? Best bloke,' Eddie said one day. 'Like a brother; a real mate.'

Ted had spent many years as a young bloke working as a farm butcher on sheep properties, until he decided that his passion was working with cattle, and he found work on cattle stations across the Northern Territory. These days Ted is retired and, with his wife Joan, has settled on a few acres with some sheep to keep the grass down. I don't need too many excuses to call into their farm to say hello. If Ted and Joan are at home, we sit out in the backyard to chat in the shade of their grand old poinciana tree, snacking on Joan's homemade ginger biscuits, washed down with a cup of tea poured the old-fashioned way – from a teapot.

'I don't know for sure, but I think two of our lambs have had pneumonia,' I shouted out a little breathlessly as Ted strolled up from his sheds to greet me. 'They both died.' I sat on the tailgate of my ute and looked at him questioningly. 'What should I have done? It's not a tick or snake bite, and from what I read on Google, it seems the lambs had pneumonia.'

Ted saw my worried face and looked at me in silence for a few moments.

'So, what symptoms did you notice?'

The symptoms? I was a bit vague as I tried to remember what I thought I'd observed. I had seen both lambs lagging behind the other lambs and making heavy weather of the trip up the hill.

'Lethargy and fever in young lambs can have several causes,' Ted said, his thin face set in its usual thoughtful expression. 'And there'll always be the occasional unexplained death. You can chat to your vet, Harri-Henry, but a vet will want to do a physical examination on the live lamb. Or if it's too late, do a post mortem to establish the actual cause. If you're right and it is pneumonia, at the end of the day, even with treatment, chances are the lambs would die.'

He was right of course. I had to agree it wasn't financially viable to involve our vet for the sake of two little lambs. Even really cute ones. I was beginning to feel quite dejected as Ted reached out and gave me a comforting pat on the shoulder.

'Ah, well, that's the thing, you see. There's no magic formula for stopping pneumonia, and our summer heat waves and high humidity, and the never-ending dust, all contribute. Just make sure your lambs don't sleep on damp bedding, and they'll need good ventilation in the pens.'

I was pretty sure of my 'lamb pneumonia' Google diagnosis, and a brief phone chat to Meggs next day confirmed this. She agreed with Ted about the need for diagnostic efforts and pathology examinations just to be sure. And if pneumonia was confirmed? Well, this is an infectious disease in young lambs and most lambs are exposed to this. The antibodies in colostrum help control the infection, but sometimes insufficient colostrum, or weakness in the lamb as the result of a difficult birth, are underlying issues.

Treatment would require antibiotics such as penicillin, and in serious outbreaks all exposed lambs would be treated with antibiotics for several days. There is a vaccination available called nasal IBR-PI-3 vaccine, given to baby lambs at two or three days of age, to help reduce problems in some cases. Treatment of the ewe flock with sulphonamides prior to lambing can also help in problem flocks, with routine treatment of all lambs with a long-acting antibiotic when an outbreak occurs. All worthwhile and valuable information. It was something for me to think about, but it would be hard to justify this extra expense. And, in the end, I decided to do nothing.

Well, it was a rotten week. Two days later, I found yet another sick lamb and a mixture of dread and disbelief flooded through me. Surely not again. This was just about the last straw, as far as I was concerned. I tried not to look on the gloomy side as the poor sick lamb lay down in an awful state. The sheep were sleeping out in the lovely open night paddock but our weather at the time was so hot and humid, and the wind was dusty. It just kind of sucked.

Next day, prepared for the worst, I snuck out early and peered into the sheep yards to find a sick lamb standing in the corner. At least he was still alive. The foreboding in my mind continued for days, until at last it seemed this miserable little fellow was going to rally, and I cheered up considerably. To my delight, he survived and grew into a rather small but very nice little sheep. This one nailed it! And much to my relief, the rest of the lambs stayed in good health. Over the years since, we have never ever had a recurrence of this dreadful rapid illness, the one I self-diagnosed on Google, as lamb pneumonia.

Afterbirth Issues and Other Stuff

It so happened one day that I came across long bits of stringy flesh hanging out from the butt of one of our ewes. I stared at it for a moment or two: this was something I hadn't seen before and I was wondering what I should do about it. I had my own misgivings as I continued to ponder this state of affairs, guessing it was an afterbirth issue, when my pocket phone buzzed and Sally rang in for a chat. Sally and Ben are almost neighbours and they farm sheep and cattle just a bit further down our road.

Sally is the farmer and Ben manages a small livestock transport business. I'd heard that he's able to skilfully reverse his small semi to feather touch loading ramps of any size, as he picks up cattle and horses from farms round about. Sally is a ball of energy, practical, and always up for a chat, and we've become good friends. She's always willing to listen to my stories, and as I began to ramble on about the bits of stringy flesh that were drooping from the sheep, Sally said she was fairly sure that this would be the afterbirth – or placenta.

'It may not always drop by itself,' she told me. 'You need to cut off the piece at risk of being trampled on by the ewe. Don't cut it too short and don't try to pull it out – you can cause excessive bleeding if you do.'

And it wasn't just that. Sally was at pains to point out the importance of always checking the afterbirth, which is supposed to be all cleared up three or four days after the ewe has given birth. If not, she advised me to ring the vet. From this conversation, I learnt that afterbirth issues were more complicated than I'd realised. Put very simply, there could be ongoing issues with metritis, or infection of the uterus, apparently quite common after a difficult delivery.

I thanked Sally for her suggestion and I went back to cut the low hanging placenta at the level of the hocks. For the next few days I monitored the ewe for any possible ongoing concerns. But she healed up well and to my relief I have never again come across this issue in our sheep. Everything plops out at the right consistency and within the recommended period of time.

Bottom's View – A Prolapse?

Learning how to manage a prolapsed sheep is the most unpleasant job of all. A vaginal prolapse is a significant welfare issue for the affected ewe, and most commonly occurs at the onset of lambing in heavily pregnant ewes. The vagina, sometimes up to and including the cervix, is pushed through the vulva and ends up outside the body. The internet readily presents a range of horrible details and treatment plans and, without treatment, a prolapse is invariably fatal. Even with treatment, a proportion of ewes will not survive.

A prolapse was something I feared, and sure enough, one day out in the far paddock ... ah well, there it was! The massive protuberance was daunting and I felt tempted to yell, 'Phone the vet!' Although a vet with an epidural and sterile technique has a better success rate, this choice is not normally an option for a small-time farmer. It is just not economically viable.

To make matters worse, I was at the local store that same morning, picking up the papers, and in conversation with Uncle Harold James I mentioned the fact that I had our first prolapse to deal with.

'You'll have to shoot her,' his matter-of-fact voice pronounced unhelpfully.

'Really?' I said blankly. 'But why?'

Chapter Seven: Getting into the Nitty Gritty of Animal Management

'If it happens once, it'll happen again.'

I was a bit alarmed. 'That's not good.'

I most certainly didn't relish the idea of just getting rid of her. Uncle Harold had his own ideas about managing sheep and wasn't afraid to express them. He could see I was still totally unconvinced and went on about various related issues – until I looked him straight in the eye and told him to 'Get stuffed!'

But I was totally unprepared for the awfulness of a prolapse. Covered in muck and hay, the huge organ hangs out like a giant melon, while the sheep's poor face says it all, 'Bugger!'

The first time we tried to 'fix' a prolapsed ewe it seemed that we would never win the battle. I watched in awe as The Engineer skilfully manipulated the organ back in, and as he did so, the straining ewe was finally able to have a really good pee. This is not possible when the organ bulges and blocks the passageway. But then just as we began to get excited about the apparent success, the organ oozed back out and the sheep was miserable once more. The Engineer suggested that it might be easier for me to do it – with my girly hands. But the very idea of me having to push this huge thing back inside the sheep was '?#@*%?'.

'Tell me again,' I said to The Engineer. 'Why are we farming sheep?'

'It seemed a good idea at the time,' he answered cheerfully.

Discussing prolapse issues with Meggs helped enormously. Meggs said to wash the dirt and shit off first with warm soapy water and disinfectant; then dry off the organ with paper towels; and then liberally sprinkle table sugar all over the extended tissue and the sheep's bottom. It seems the sugar shrinks and dries the area concerned and can help prevent a prolapse happening again.

As for the pushing bit; I was to keep pushing until my arm was nearly to the elbow, and the mass of prolapsed organ went back inside – and stayed inside.

'It should just sort of click into place,' Meggs said.

The first time I did it myself, I don't know how I did it: pushing, resting, pushing again, as both the sheep and I emitted long drawn-out groans. I really had to give it a shove, but every time I pushed it in, the whole mass oozed back out again. I could hear my session with Meggs going around in my head.

'Push until your arm is well in and everything should just sort of click into place.'

I still have no idea what this really means. But I must say it feels amazing when everything does disappear out of sight, and actually stays inside the sheep. This is a serious problem, and pushing and prodding the swollen mass back in does seem almost brutal, but it is a lifesaver.

Stumpy showed us an old-fashioned and effective way of immobilising sheep that we now use if any of our sheep have a prolapse. Using a straw bale, we position the sheep on it, and in its simplest form, this anchors her so she can't run away. This gives us the force of gravity in trying to work the prolapsed organ back in.

Earthy Murphy said that we should try his way of stopping a recurring prolapse – a method often used by old-timers out in the bush, using four pieces of handy hay baling twine, all cobbled together with some fancy knot work. Earthy Murphy cheerfully explained to me with the utmost patience, what to do.

'First, ya restrain the ewe to stop her wandering off,' he said. 'This is easy. Then ya tie the baling twine pieces together, centre them on her shoulders at the back of her neck, cross them across her breastbone, run the two ends under her armpits, cross them again over her back, run the ends under her hind leg pits, alongside her tail, and along her spine to tie tightly to the original twine across her shoulders. The twine needs to be tight enough to keep her back slightly arched so that she can't use her stomach muscles to push. Then ya tie short pieces of twine above and below her vaginal opening to keep the prolapse in place. Call me anytime! I can show ya!'

'OK. Thanks.' I nodded dumbly. This may have worked for Earthy Murphy and his bush mates, but I had no idea what he was on about!

When I bumped into Sally at our local shops, I mentioned that we were having some prolapse issues and told her about Earthy Murphy's

Chapter Seven: Getting into the Nitty Gritty of Animal Management

hay-bale twine solution. She stared at me for a moment or two, and then she began to laugh.

'So, he told you that old story! Shit no! Buy a harness. A ready-made one.' Her grey eyes shone with enthusiasm. 'It's a game-changer. Humane, effective – and will stop the ewe pushing but won't stop her lambing.' I could see her still chuckling at the twine idea as she wandered off to finish her shopping.

Well, it didn't take me long to go online and find the farm shop website, and I didn't think twice about purchasing not one, but two handy prolapse harnesses. It was just a few days later that Derek delivered the parcel while The Engineer and I were at the gate chatting to Eddie. I unwrapped the package and proudly showed them my new purchase for the farm.

'Strewth! What's this for?' Eddie asked in a surprised tone of voice as he picked up the pieces of webbing.

'It's supposed to stop a recurring prolapse and keep the organ inside the ewe where it belongs,' I pointed out. 'That's the idea of course.'

'Dunno! All news to me,' Eddie muttered as he began to weave and rearrange bits of webbing through and around the buckles.

'How's it supposed to work?' The Engineer pondered as he attempted to read the paper instructions.

I rang Sally for some fine-tuning pointers and she wasn't shy in giving her assessment. 'Goes over her arse,' she bellowed over the phone.

It must have been nearly three weeks later that we had another ewe present with a prolapse, and after struggling a bit to get the organ back inside, we managed to fit the prolapse harness on her. And as soon as the harness was clipped on, it was instant relief for the ewe. Dressed in her harness she came across as a clumsily wrapped woolly box, with her butt now free for vital things like peeing! And having babies. I told her that 'it's all good' and when she has successfully delivered her lamb, she will thank us for it.

The prolapse harness is a simple fix, and for the troubled ewe it's a life-changing purchase. It may seem a real hassle, especially if we have to refit it a number of times over a number of weeks, but admiring the proud sheep with her new healthy baby after delivery is such a rewarding return on our efforts. To date, we've experienced a number of healthy lambs successfully born with sheep wearing a prolapse harness.

Why Do Lambs Sometimes Die?

If there's one indisputable truth for a farmer, it's this: sometimes animals die. Life and death are part of the work on a farm and I've grudgingly come to terms with this. The painful truth is that sometimes we can prevent this from happening, and sometimes we can't. Seemingly healthy animals get sick. Some get better but some don't. We don't very often call our vet. In practical terms, the cost will be several times the worth of the lamb – and too often the animal will die anyway.

Sometimes, in spite of every effort to save a mother or a baby during lambing, the farmer loses. Exposure in wet and cold conditions can rapidly kill lambs, and twins and triplets are particularly vulnerable. And sadly, I can't assume every ewe will be a good mother and care for her new baby. For instance, lambs can suffocate if membranes around them have not broken at birth and the ewe has not licked the face well.

We keep our sheep penned up in yards and barns at night for protection from predators. But it was some time before I understood the value of having sheep yards and pens large enough to give ewes the space they need, to get away from the flock in order to give birth. One morning I discovered twins had been born during the night: one was alive but the other had been trampled on, and I found it most distressing, dealing with the tiny lifeless baby lying dirty on the ground. Since then we have built bigger yards and barns with wide lanes to manage our growing flock of sheep, providing space for the ewe to protect her new baby from pushy sheep and inquisitive lambs.

Things To Do For Best Farm Practice

1. Weaning

Weaning can be stressful to lambs and may result in a lower weight gain, decreased immune system functions and poor animal health, which may then lead to increased vulnerability to disease and infection. Our preference is to let nature run its course and allow animals to wean naturally. Natural weaning begins as soon as the lamb's intake of natural grass and hay increases. At the start, the lamb's eagerness for milk ensures them a steady milk supply and this contact between mum and babies helps to maintain a high milk production, but after we let them out into the grassy paddocks, we have seen the ewes walk off during suckling sessions, marking the start of a very gradual weaning process.

2. Tail docking

I'm the first to admit that, as first-time hobby farmers, we had to learn all about tail docking. I consulted Stumpy. 'Does it hurt?' I'd been thinking about this question for a while.

He looked me up and down and said, 'Not much.'

I went with this. Blowfly strike is one of the most significant health and welfare concerns for sheep in Australia and tail docking in sheep is a highly effective preventative technique, markedly reducing the incidence rate. Blowfly strike is nasty once the maggots take hold in a soiled patch of wool, having the capability to creep into the flesh of the sheep, causing huge pain. If left untreated, the sheep will die.

Ringing lamb tails is the easiest and most common method for tail docking and, in our experience, the lambs feel only a little discomfit for a little time. We ring tails when the lambs are a bit over a week old and look strong and healthy. Using an elastrator tool, a special rubber ring is applied to the tail. This then slowly stops the circulation and the tail falls off in seven to ten days. The recommendation for tail length, according to Google, Stumpy and Meggs, is the third palpable joint of the tail – which is about the length that just covers the vulva in ewes.

3. Newborn lambs with a raw, chewed tail

It is rather alarming to find a newborn lamb with a raw, chewed tail. Why it happens seems to be anyone's guess. My own theory is that on occasions some ewes get carried away, obsessive even, when biting the umbilical cord and just keep on chewing other hanging bits – the tail in particular. We've experienced this unusual behaviour on a few occasions, especially when the sheep has given birth to twins.

The first time I caught sight of a chewed tail, it was in a disgusting state. And it didn't look much better after we sprayed the stump with antibacterial and insect repellent wound spray – in bright pink and purple colours. This practical treatment is a good simple solution; and if the chewed tail is long enough, we will also ring it at the recommended third palpable joint.

4. Worming

Internal parasites are a common health problem for grazing livestock, especially sheep and goats. There are a number of effective management practices to lessen the risk, e.g. antiparasitic drugs that expel parasitic worms and other internal parasites from the body. In Queensland, Ivomec® and Cydectin are considered to be most effective, although there is a continuing argument about whether we worm too much. Or not enough. Worming reports are emerging, stating that treating every sheep in a flock is expensive and ineffective, and leads to accelerated parasite resistance.

Some old farmers take this quite literally. We have a new near neighbour, Tony, who lives two farms away from us. A casual 'she'll be right' sort of bloke. 'You only need to worm sheep when their gums are white.'

Seriously? Tony only bothered to treat the animals that were indeed suffering and I worried about his sheep. It was dispiriting to realise quite a few had died from that kind of thinking.

Sally suggested the worthwhile exercise of a faecal egg count reduction test. To do this test, a sample of fresh sheep poo from the ground is provided to a vet for worm counts. The test will compare faecal samples taken before treatment with one taken 10 to 14 days after treatment. If the drug is effective, egg counts should be reduced by 95 per cent. Samples can be sent direct to laboratories to have egg counts done, but it is advisable to get expert advice to help interpret egg counts. The results will help determine which drench is to be the most effective.

5. Vaccinations

Vaccinations play an important part in keeping sheep healthy. We vaccinate our sheep with an Ultravac vaccine to prevent major sheep diseases with fearsome names such as pulpy kidney disease, black disease and malignant oedema, as well as blackleg, swelled head in rams, and tetanus. If the ewe was vaccinated before lambing and the lamb received colostrum, then vaccination is not necessary until the lamb is three months of age.

6. Castration

I did become a little preoccupied about doing the right thing as I considered whether to castrate our ram lambs or leave them entire. I needed to understand why it can be necessary, so I did my Google thing, reading various vet and farm sites, and I learnt that castration is performed for management reasons, as well as perceived benefits to meat quality. The general opinion is that wether (or castrated) lambs are easier and safer to handle and manage, and less likely to fight, having reduced aggression and sexual activity.

During our first year farming sheep, we castrated all the ram lambs. But then we noticed that the lambs, at around ten months of age, seemed rather small and insubstantial in body structure. So, not convinced to continue with the castration theory, I decided, understandably, to consult some local advice.

On my way through Woolworths one afternoon, I bumped into Ted Graham deep in conversation with Eddie, in one of the aisles. Their banter was rich with references to cattle and sales prices and, of course, the weather.

'Tell me about castration,' I asked eventually. 'How much of an issue is meat quality when our market lambs have a life expectancy of nine to ten months?'

Being a retired farm butcher, Ted was just the man to ask. He was of the opinion that there was no difference in the taste or tenderness of the meat from a young ram lamb as opposed to a wether or ewe lamb. Always supportive with a comment or two, Eddie agreed. And so, I couldn't help but question why I would consider castration, if the meat tastes the same and ram lambs are quite manageable.

The following year we decided to leave the little ram lambs entire to see what would happen and it all turned out for the best. Over the months, I scrutinised their progress and by the time the ram lambs had reached ten months of age, it was quite evident that they were considerably bigger, bulkier and stronger than the castrated lambs of a similar age, grown in our first year.

There was another thing too. Separating lambs for breeding or for market is much easier when the boys keep their testicles and tails. Tails are also left on the market ewe lambs. A lamb with a tail goes in one direction; a lamb without a tail is kept for breeding.

Southdown sheep are a gentle breed, but for ease of management, when the ram lambs are big, boisterous and at the point of making noticeable and unwelcome advances to our breeding ewes, we do find it best to separate them into other small paddocks on the far side of our property. The only in-house fighting I have ever observed has been from ewes in season, head-butting each other as they bounce off down the hills.

7. Sheep shearing

Most sheep breeds will need shearing for their comfort and well-being, and to prevent a number of problems which can occur if a sheep's fleece grows too long. At its worst, the excess wool can cause overheating. Urine, faeces and other bits and pieces may also become trapped in the wool, attracting flies (resulting in maggots) and other pests. There's a fair chance that this will develop irritations and infections, and endanger the health of the sheep.

Wally is our shearer and he likes to remind me that when he was 'just a lad' he could shear almost two hundred sheep in a day; this seems a lot of sheep! He trims the hooves of our sheep and goats as he shears the sheep, and the wool considered worth keeping is packed into giant bags, ready to be sent off to some local spinners. This arrangement saves us the trouble of having to dispose of the wool. What I still find fascinating is how my old dry hands become beautifully soft after a morning handling lanolin-saturated wool.

Our small-scale shearing facilities are basic – a clean, dry barn with wooden floors, simple holding pens, and lanes for flow-through management. Pre-shearing, we keep the sheep from feeding the night before, however water is available. Wally tells me that this keeps them quieter and more comfortable during the shearing process. He showed us how the sooky, quiet sheep keep their back legs straight and relaxed, and these are the ones that are quick and easy to shear. But the sheep that bunch their back legs up – ready for a fight – are the exuberant, feisty ones. 'A bit like a boxer curling his fists as he eyeballs the enemy,' Wally chuckled out loud.

Wally brings all his own gear, including an electric clipper machine with hand-piece and combs. My job is to work Belle, our working Border Collie, in the yards, keeping a steady line-up of sheep moving forward. As the sheep lose their fleece, we can see now how well the sheep have been doing, which ones are in top condition – those as fat as butter – and which of the older ewes are on the skinny side.

Lice irritate sheep, causing them to bite, scratch and rub on trees and fences, and so The Engineer applies a lice treatment in the form of a backline application down the back of each sheep as it leaves the shearing yards. It's quick, easy and effective. Our last chore of the morning is sweeping and picking up the daggy, loose, unwanted bits of wool. Then comes our favourite part of shearing: sharing coffee and cake with Wally.

8. Teeth

Sheep only have teeth on the bottom jaw. A lamb has sharp little baby teeth but at about a year old the two central teeth change into fatter white teeth. The following year the ones on either side of the central two change to adult teeth. Three years after this, the whole mouth is in its adult form – a row of broad white teeth. As the sheep ages, the teeth get longer and start to weaken, with gaps forming. Soon the remaining teeth become wobbly and fall out, although they can actually graze with no teeth at all.

An early morning song from English Fred our Southdown ram

The miracle of a new lamb being born

Nearly there, Ma

The ewe licking one of her new born twin boys

A sleepy baby lamb in a nest of lucerne hay

Mothers and babies

Some baby lambs off for their early morning 'hoon'

These two lambs were born at the same time which was a little confusing for their mothers – and me!

A shady tree is a must for summer shade

Lambs at play as the sun goes down

Little Boris cuddled up beside my shoes

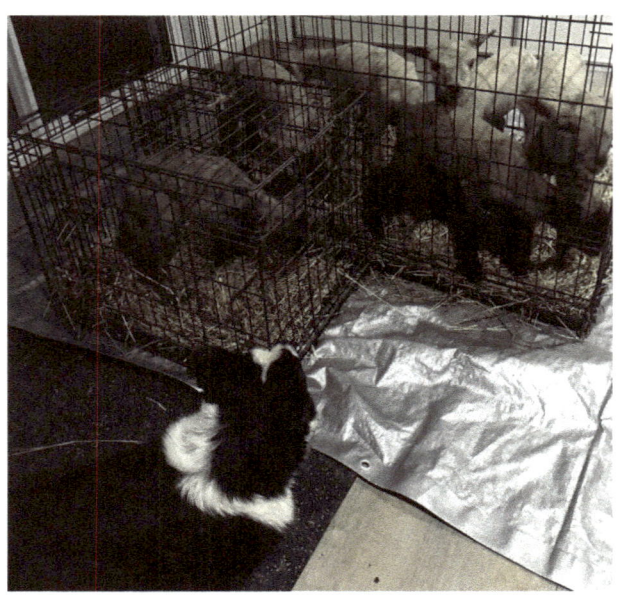

Belle watching over the poddy-lambs in their dog-crate houses

A lamb with a 'chewed tail' which has been sprayed with pink antiseptic spray

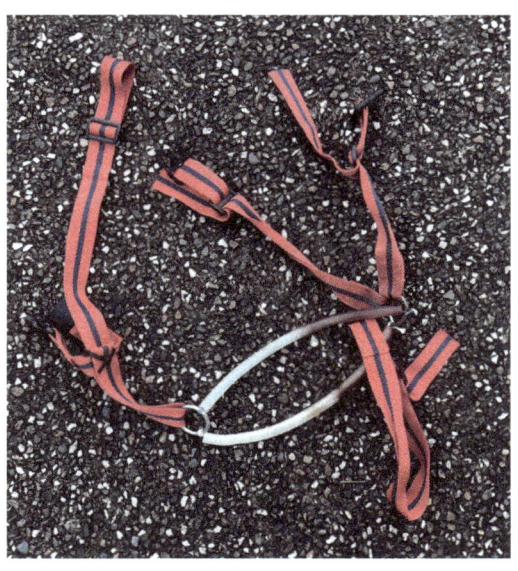

A real-life prolapse harness

A sheep wearing her prolapse harness with her newborn lamb

A sheep wearing her prolapse harness

EIGHT

The Farmer and Her Dogs

There is something about working with a dog that I find irresistible, especially the warm bond of love and shared dependency that grows stronger over the years. My passion in life is training our German Shepherds to compete in dog obedience trials. We train hard and compete well and there is nothing like the feeling of exhilaration when we achieve the highest level of obedience and top scores at State and National competitions.

But … anybody who has farm animals will appreciate my interest in working farm dogs. In my imagination, I'm the perfect shepherd with a perfect sheepdog, striding confidently over the hills herding sheep.

I mentioned to Charlene I was thinking about getting a working sheepdog. We were sipping frappés in the courtyard of the café at the time, and she nodded happily. Charlene has lived in this area for a long time and has the reputation in the district of knowing everyone. It was plain that she viewed my project with enthusiasm and her response was immediate. 'Bluey. Talk to Bluey. That old codger – he's got all sorts of working dogs. Stumpy Tails, Kelpies, Border Collies and Blue Cattle Dogs.'

It was only a few days later that Charlene drove me out to meet up with Bluey. Bluey is a friendly, old-fashioned bloke with an interest in breeding working dogs. He has spent half his life out west in the hard country, but these days, lives on acreage the other side of the range with his wife Thelma and old Aunt Ann. As we parked Charlene's ute at the front of the house, Bluey greeted us enthusiastically. I found him immensely likeable with his broad grin and unpredictable sense of humour.

As he shook hands with me, he bellowed, 'Ha! The old sheila who has the balls to be a farmer!' And he roared with laughter. I pretended that I hadn't heard this one before. But Thelma scolded him for being a toe-curling total embarrassment, as she produced a scrumptious morning tea – real coffee and hot buttered scones.

Everything about their little cottage is old. The high ceilings are cracked and the side veranda walls look decidedly rickety. From the kitchen, French doors open out onto a wooden deck, protected from the western sun by two enormous bougainvillea shrubs and a jacaranda tree. The farmhouse kitchen is rough but comfortable, with an old wooden table laden with odds and ends, and there is an outdoor dunny* just around the corner.

Chapter Eight: The Farmer and Her Dogs

I felt totally at ease, chatting effortlessly about our farm and our animals and why I thought a Border Collie might be a good idea. Aunt Ann had been a shearer's cook as a young woman and had since spent many years working on sheep properties with Border Collies. Her old face crinkled into a wide grin, as she leaned towards me and said, 'They be best dogs, my girl. Yep, they be best dogs indeed.'

Thelma's passion was breeding and training Border Collies and her dogs consistently won at local sheepdog competitions and sometimes at national trials. There were hanging pictures of Border Collies in the living room and down the passage way, and when Thelma saw my interest in the dogs, she began to rummage through the pockets of her jeans to find her phone, in order to show me even more photos of her past and present Ekka** winners. I put on my reading glasses and saw more Border Collies, all colours and shapes, sitting proudly behind a selection of over-sized trophies.

After morning tea, Thelma invited me out to the kennel runs to see the dogs. There was one extra-large enclosure with Border Collie puppies romping and rolling around together on the grass. And they were gorgeous. Fluffy black and white balls getting under our feet and trying to climb into my lap. All ten of them. All at once. Spellbound, I could have stayed there all day. I was in love. Which was a pity because Thelma explained that all these pups had been sold in advance. I was profoundly disappointed and found it difficult to get the puppies out of my mind.

Thelma went on to suggest that good working sheepdog pups are often to be found on rural classified websites. I looked at her doubtfully: it was a good idea, but would I know a good pup from a not-so-good pup?

We talked at length about my farm and what I wanted a dog to do. And how a Border Collie would fit in with our household of German Shepherds and a Belgian Malinois. We went into the details of how to choose a good puppy and manage the training required for a dog to herd sheep.

'A good working Border Collie pup,' Thelma explained, 'has the instinct from birth. As soon as they can walk, they show interest in rounding up chooks and ducks – even cattle and sheep if given the opportunity. This is a partnership where a job must get done and Border Collies have been doing this job for years. No person can replace the skill of a Border Collie. These dogs don't need to be taught how to herd, but they do need to be taught where and what to herd. How to drive, how to lead and how to work the sheep yards and race. A good dog with a bit of bark and bite can work in all weathers.'

I went home in a thoughtful state. Over the next few days, I resorted to Google to search for information about Border Collies and to look at herding dogs for sale. I studied lots of training YouTube clips on how to cleverly control the herding instinct. I learnt that working dogs are one of nature's true wonders, and if you're searching for the ideal partner for stock management, then to look no further than a Border Collie.

And so, it wasn't too long before Belle, a five-month-old working Border Collie pup from a sheep farm in Victoria, became a member of our gang. We loved her from the start. This new pup, with her mischievous bright eyes, black and white face, and one ear sticking up and the other flopped over, was the complete cute package of comical charm.

As expected, learning how to train Belle, with her bitey face, waggly tail and alert, intelligent, exuberant personality, was an art in itself. I read books about sheep herding. And I listened to recommendations from sheep herding blokes, hanging onto every bit of advice they offered. I watched so many YouTube wonders showing the best Border Collies in the UK that I was convinced I knew it all.

One morning, prepared with a long rope line and a piece of poly-pipe to represent a real shepherd's crook, Belle and I began our herding journey. She learnt to hit the ground fast when I bellowed, 'Stop!' At first, I rewarded her with a piece of cheese but it wasn't too long before this wasn't necessary. For Belle, with her innate instinct to chase, fetch, drive and hold animals in position, herding sheep is what life is all about. And the sheer fun of being with the sheep is all the reward she needs.

There's an old saying: 'A good shepherd can make a dog'. But in my experience, it is a case of a good dog making me into an 'all right' kind of shepherd. And the best part is that, sometimes together, we

Chapter Eight: The Farmer and Her Dogs

look amazing. Other times it is just a shambles, with sheep scattered and stressed – and me frantic and stressed – while Belle, with endearing larrikin energy, hoons around having a wonderful time.

My commands are not actually conventional. I was fairly confident that I could master the authentic commands of *Come-bye* and *Away*, but more often than not my directives fall back into more desperate bellows of *What the hell are ya doing?* and *Get your hairy arse over here.*

Belle works for the sheer love of it, and has proved to be invaluable in helping us move and manage sheep around our farm; it's hard to imagine a person being so efficient. I have mostly curbed her habit of biting the sheep, and I'm overwhelmingly impressed by how she will stand her ground with fractious ewes and can herd the cows, including Big Ernie the bull, into places they don't want to be.

She's learnt to wait for that wave of my arm. With her body down low, eyes fixed intently on the grazing animals, she's ready to streak across the paddocks with only the white tip of her tail, like a wild flag, visible in the distance. She will tear after a stray sheep, return them to the mob, and turn her happy, panting face cheekily to me to ask, 'I'm here to do a job so what's next?'

After I'd spent countless hours training Belle, our daily herding efforts were good enough, I thought, to post as videos on Facebook and YouTube. It didn't seem a bad idea to use social media to show off our *awesomeness* as a 'shepherd and dog team'. Well, nothing much was said in regard to my video clips until one day, Keith, his face quite expressionless, questioned me about my Facebook posts and challenged me to give a 'herding demonstration'.

'Really?' I was a bit surprised and I felt myself blushing as I stared at him. Keith is a country bloke, lanky and comparatively young, with an air of effortless command. The expression on his face showed that he thought I knew nothing at all about working dogs and herding sheep. In his own eyes, he is a fountain of knowledge and therefore considers it necessary to impart his wisdom to anybody who crosses his path. And I was a gift!

Keith believes he knows most things about sheep herding, despite the fact that most of his herding experience is based on the occasional chat with old-timers competing at the Ekka, and impressions and information gleaned from YouTube clips – not mine, it seemed. He owns three dogs – tough, rowdy Kelpies – and his way of handling cattle is to zoom around his farm on a noisy trail bike. With a lot of yelling and rushing about, Keith and his dogs inevitably scatter cows in every direction.

I couldn't escape the feeling that this challenge was *not* a good idea. I was, to say the least, a little wary of committing myself to any action and I did my best not to think about it. I really was hoping that the whole idea would disappear. Well, nothing more was said for ages and my fears were beginning to fade away, then Eddie casually spoke to Scooter, who then phoned me.

'Now then, Harri-Henry, when's your herding show going to happen?' He chuckled. 'Unless you don't think you're up for the challenge.'

'No, no! Challenge duly noted. I'm up for the challenge. Count me in.'

A slightly nervous expression flitted across my face and my heart sank. I didn't like it a bit. But it was going to be no use trying to put up some feeble resistance, and so with great charm and studied carelessness, I waffled on about being rather new to herding sheep, but I could, at a pinch, demonstrate our current level of expertise on the following weekend.

After Scooter hung up, I sat for a few minutes to think. My misgivings about letting others watch Belle and me herding sheep were very real. I had to admit, I could imagine Keith's lofty enjoyment if Belle did her own thing and then, between us, we made a right mess of it.

So there and then, I decided to invite my own support team of dog-obedience friends to be in attendance and cheer us along. So that was sorted.

And sure enough, they all came. They were in a rowdy, festive mood and found plenty to cheer about. They cheered wildly when Belle pranced into the yards, and then cheered me when I picked up my favourite piece of poly-pipe shepherd's stick. And they cheered when Scooter, Eddie, George and Dusty – and an unbelievably big turn-out of folk from the Bull & Barrel Country Pub – all rolled up and perched themselves on our cattle rails. I looked around at everyone in some surprise. I hadn't realised so many spectators were going to be in place.

'It's the real deal. It's not often we get to see a real working dog in action!' Dusty told me.

From their position on the rails, everyone had a great view of the back fence and the paddocks where the sheep were busy grazing. I was not feeling at all confident. Keith, his hands deep in his pockets, eyed Belle with his usual deadpan expression. He was completely unruffled as she cavorted and fooled around with the onlookers, as they chatted amongst themselves.

I decided to boldly describe what I thought should happen, and my intended objectives, in some detail. The plan? Belle was to 'go back' to the bottom paddock; herd the sheep from behind; drive them up through the first gate; through the 'new' paddock; through a gate into the middle paddock; and up into the sheep yards. I really was hoping that this would easily demonstrate our competency and sheer skill as a cool dog/shepherd team. I whispered in Belle's ear and literally begged her to put on a performance that would do credit to a real working Border Collie in action.

But – it was not to be.

She began well, and her wide 'outrun' was quite remarkable. But things went from bad to worse after this. Indeed, it was soon very clear that the demonstration was not going quite as well as I hoped. Seemingly quite deaf to my shrieks, Belle, with her unmistakeably eager face, managed to disgrace herself by eyeballing some little lambs until they broke free and raced off down the hill in the opposite direction – quite some distance from their mothers. I glared at her helplessly, and hands on hips in despair, tried not to swear. To say I bellowed was an understatement, and howls of laughter arose from my support team leaning over the middle fence line, where they had a grand view of Belle. They egged her on with helpful advice and suggestions.

'Hey Belle! Turn on your GPS!'

'On ya, Belle Baby! Away! Come-bye.'

'You what! It's the other way. Go the other way!'

'Hoy. Other gate, mate!'

These comments caused considerable hilarity and I could hear peals of delighted laughter as they clapped their hands in ready encouragement. This was more than quiet leg pulling. Scooter was just as bad. I heard his deep voice roar with laughter.

'Hold a mo', Belle, I need a selfie!' Then they all decided to take selfies, with Belle in the background racing away and chasing the lambs.

Keith enjoyed every moment. With an amused little smile on his face, he made an impressive figure as he strode along. The most exasperating thing was the way he kept clearing his throat and going 'harrumph' as Belle finally chased the lambs around to meet up with the unhappy ewes. It was an embarrassment. I'm quite convinced those lambs had smirks on their woolly faces as they dutifully followed their mothers up into the yards.

It was over. I leaned on the gate for a moment, enjoying the berserk cheers of my friends giving their anticipated standing ovation, and I turned around and unashamedly gave my audience a thumbs-up sign. Keith, with a snort of laughter, gave a small nod in appreciation. The annoying part was that now I had actually closed the gate on the sheep, he narrowed his eyes and began to speak of the old days when herding sheep was left to the professionals. I couldn't say anything. He was a difficult man to impress. But it did seem appropriate to throw my cap in the air. What did it matter? I was happy. And, for sure, Belle was happy.

I wasn't too disappointed with our lack of success. But then, off the top of my head, I decided we needed an unplanned fitting finale – chook herding.

I invited everyone to stroll through the gardens to watch, as Belle, with style and flair, began to herd eleven silly silkies through two groups of complaining guinea fowl, through the chook-yard gates and into the chook house. She was careful and methodical. When the hens were all put away and I was just about to close the gate, Belle darted back inside the hen house and came trotting back a few moments later, gently holding a large guinea fowl egg in her mouth. She trotted past me and over to Keith, where she gently plonked the egg at his feet. Her eyes were laughing up at him. A slightly confused expression wavered on his face as he drew his brows down in a deep frown. Then he said, 'Oh.'

There was a moment's silence – then everyone laughed. 'Looks like everybody's kicked a goal,' Scooter grinned. And Dusty said afterwards, 'For sure that Keith, he's just a tiresome think bucket. No idea what he's banging on about.'

I couldn't help but laugh. Everyone stayed on for a sausage sizzle. The Engineer cooked his favourite curry snags and onions on the barbeque and Keith had his recently acquired egg, fried on both sides. He seemed to enjoy it.

Funnily enough, after this episode, Keith and I got on rather well. 'Yeah,' he admitted as he climbed up into his truck. 'Yeah, not a bad working dog.'

Hanna's Stay in Hospital

It was perhaps wishful thinking, but taking our sick and sad Malinois to the vet, and hoping that it wouldn't cost much, was a bit of a stretch of the imagination. Who was I kidding? Hanna had vomited up most of my red gumboots, and this made me really cross. And then really concerned.

She did look sick as we went through to the consulting room where Meggs bent down to examine her. Hanna, trembling and anxious, stared up at me as I watched Meggs do a thorough examination. Then regarding me unsmilingly, Meggs became very brisk as she made a list of the tests that would need to be run. 'Because,' she said, 'it sounds like something's going on inside.'

Leaning across the table, I didn't reply straight away. It did seem I had no choice but to leave Hanna at the surgery and my misgivings proved to be well-founded. Internal exploration would find that not only had Hanna eaten part of my boots, she had also consumed large pieces of poly-pipe and the sides of a plastic ice-cream container. Somehow this had all escaped my notice.

She looked perky enough when I picked her up four days later, but I wasn't feeling as perky when I was presented with the invoice and the accompanying jar of bits and pieces removed from her intestine and stomach. The crisis was over.

Belle, her eyes dancing with pleasure, greeted Hanna with enthusiasm. I showed The Engineer the invoice damage. He said that in his opinion, not only do dogs keep us humble, they keep us poor.

Belle's Babies

One chilly evening, I was hurrying along one of the flattened dirt paths the sheep make across the slopes, when I noticed Belle sniffing a little bundle lying under a tree. It was a baby lamb all alone in the paddock, rather damp and silent, but still alive. I scooped her up and went off to find her mother. The ewe was easy to spot as there were obvious signs she had recently given birth, but she showed no interest in the sudden appearance of her lamb. As I stood there pondering what I needed to do, Belle trotted off into the gloom and tracked down another dirt path. Somewhere nearby she had picked up a scent and led me to another woebegone little lamb, half-hidden in the long grass.

I simply had no idea that two little lambs had been abandoned in the big paddock, and without Belle's awareness, they would have died from cold and neglect. I felt a twinge of anxiety. What were the chances of survival for these damp, cold and hungry lambs?

The Engineer scooped them up and rushed them inside to the warmth of our air conditioner, and when they were both warm and dry, we returned them to the ewe for a first feed of colostrum. The ewe did accept her twins and let them feed, and they did survive the night. But next morning, they looked listless and felt cold when I picked them up, and their little tummies appeared hunched. I worried that they were not getting sufficient milk and it seemed easier to bottle-feed supplements for a few days, until it became apparent that the ewe was managing to produce an adequate amount of milk, and was allowing the lambs to suckle her teats. It was a special day when I knew the twins were going to thrive without supplementary bottles. They looked warm and plump as they frisked around the paddock. Relieved, I packed away the baby bottles once more.

dunny: Australian slang for an outside toilet

**Ekka: The Brisbane Exhibition or Royal Queensland Show*

NINE

The Easy-Peasy Method of Caring for Cows

You never really own cows; you just pander to their needs. For my part, the reality of farming cows is nothing like my visions of a small herd happily grazing all year on gorgeous pastures. Nobody mentioned having to lug heavy bales of hay from the shed to the paddocks; being tired and hungry from a long day; and driving rain beating down on my cap. And, then there was the time when Scooter and some neighbours were assessing our new cows and I stumbled over a half-hidden rock and smashed headlong into a large moist cow pat in the cattle yards. So gross. It was 'Oh shit!' There was mud on my nose and cow poop on my ear. Howls of laughter greeted me as I picked myself up and wiped my face with the bottom of my t-shirt.

Of course, the question was not whether we would want cows, but what sort? What were some cow choices? I considered every breed out there eating grass; some looked amazing and it was difficult to decide. Then we bumped into Kenny outside the newsagent. He was heading home from the saleyards and invited us to come over for smoko. His farm is always a hive of activity, with usually interesting and entertaining people to meet – all qualified to discuss cows and farming in great detail.

HARRI-HENRY'S FARM

'Meet Harri-Henry, the new farmer from Cliffside Cottage.' Kenny poked his head into the hay shed and introduced me to Lockman. 'She needs all the help she can get!'

Lockman glanced around and nodded hello. But over a mug of coffee he opened up and told stories of the days when he managed a crew, pushing mobs of cattle along the stock routes between Tambo and Augathella. And about some wild journeys droving cattle through the long paddock at Charleville, chasing the good feed.

The next time I saw Lockman was when he dropped into our farm on his way to the 'office'. It seemed he had some spare time on his hands and this became a regular occurrence. I will talk about farming to anyone who'll listen, and there was no shortage of conversation as we sat around our kitchen table, debating issues about cattle yards and cattle management, and potential long-term plans for future breeding stock.

Lockman suggested we always look for cows from local farms. 'Local stock have knowledge of the winds, the smells, the sounds, even the best grasses, so they settle in best,' he said one time, as he strode unhurriedly into our cattle yards and looked around with mild interest.

Chapter Nine: The Easy-Peasy Method of Caring for Cows

Leaning his elbow on the top rail he studied our old cattle crush.* Ours was not in good shape, and Lockman suggested The Engineer should rebuild the crush with a strong head bale and install a safe vet gate.

'If you need to get a vet, it's important they can get in close and not be kicked in the guts – or worse.' He spread his hands and grinned.

We were glad of his help. Our cattle crush is now a combination of the old and the new, and is much safer for us and the cows. The original posts are very old but sturdy enough, and The Engineer has installed the new vet gate and renovated the head lock structure. This enables us to be well out of the way while worming, vaccinating and doing whatever else we need to do, that the cows won't like.

After much deliberation and discussion about cattle breeds, we opted for Belted Galloways. *Belties*, as they are known, are considered to be calm cattle and generally easy to move. I knew nothing about these cows; I just liked the way they looked, and as it turns out, these gentle medium-sized cows are exceptionally easy to manage. And rib fillet steak? It's best not to mention that.

It wasn't too long before our first cows arrived. I was sitting in the kitchen, almost on top of our wood fire, enjoying my second coffee, when The Engineer calmly mentioned that the truck was at the gate. As the cows lumbered casually down the ramp, I noticed The Engineer gazing at them with undisguised delight, but I was a little intimidated by the sheer bulk of them. What were we thinking?

And, it must be said, the learning curve for farming cows has been bewildering. You could have told me anything. Especially after I pencilled in my first question: how do we tell when a cow is in season?

Okay! It seems they jump on each other.

Our next decision was whether to attempt artificial insemination or borrow a bull in order to produce more of our own stock. The Engineer, with a grin, shrugged and said, 'How hard can it be to manage a bull?' Well, Big Ernie made his presence felt in a relatively small amount of time, and our first calves were on the way.

The invitation to help with a preg-test to confirm whether Gracie was pregnant was something else. Well, yes and no. What does this even mean? You have to put your hand *where*? I gingerly stuck out my arm and put on a long plastic glove. On the outside, my face looked like I was saying, 'Yeah, right.' But in reality, it felt like my eyeballs were on the other side of their sockets, making this a potentially unforgettable experience. The Engineer said it was one of the funniest sights he'd ever seen.

One time, Lockman brought his old dad, Stockman Jack, to visit and our conversations were filled with stories about Stockman Jack's younger days in the mid-1930s and 40s. Such a tough old guy, still recognised for his drive and organising ability, Stockman Jack has always had a love for cattle, and as a small child had helped break in horses. He told us true stories about his father, Big Jack, and Grandfather, Old Jack, who were drovers before the Second World War. His dad, Big Jack, had a contract in Sydney to move cattle down the Great Western

Chapter Nine: The Easy-Peasy Method of Caring for Cows

Highway, from the sale yards to the abattoir, between the hours of midnight and five in the morning.

But my eyes began to widen as Stockman Jack told me how, as a ten-year-old kid, he would get up at 3.00 am, two days a week, to harness up a horse and sulky. He would then drive over to where his dad was droving several hundred head of cattle through the back streets of Sydney. When Jack caught up with his dad and the cattle, his father would take the sulky back home to have his breakfast, leaving young Stockman Jack in charge. Astride his dad's large stock horse and handling the full-size stock whip, Jack would take over the responsibilities of managing up to 200 head of cattle on his own, with the able assistance of a black Kelpie called Calen, and Rex, a Blue Cattle Dog.

He recalls fondly that in the dark he hardly ever saw the dogs. They did all the hard work on their own. One dog would drive and also head the cattle, and the other dog would be on every street corner to stop the beasts turning down side roads. After his dad had finished his breakfast, he would drive the sulky back to where the cattle were still moving along, and swap places with Stockman Jack. Jack would then take the sulky back to the house to get ready for school. School wasn't important to him; nothing else mattered but his ambition to be an expert stockman. And to this day, Stockman Jack is considered a legend with a stock whip.

He still can recall the time a car sped around a corner and ploughed into the cows. All the occupants – a prominent family returning from a weekend away and driving too fast – were killed. In his words, 'The lot of them were pissed. They didn't see the mob until it was too late and they killed eight cows!'

There were many times too when I ran into Trev. Trev grew up on a beef property in outback Queensland and is superbly fit and tough. He is a skilled horseman and has enjoyed competing in both 'classic' and 'open draft' competitions at the annual Cloncurry Stockman's Challenge – one of the greatest horse events in Australia.

'Yeah, it's a good way of life. Lots of thrills and spills and you get to meet nice people.' He gave a little grin. 'I kept going because it's the best challenge for camp draft competitor; it's bloody unreal.'

But these days, Trev's swapped his passion for camp drafting and bronc riding for team roping. 'It's less impact,' he told me. 'My bones can only take so much; I know that and I've given them a hell of a hiding all my life.'

Trev has set up a metal calf, with horns and a big head, to squat out on the grassy oval behind the Bull & Barrel Pub. This is for the kids' game, *Roping the Calf*. It's really a contest, and the aim of the game is to swing a lasso over and around your head, then swing the rope out to reach over the metal head and horns. One late summer's afternoon, The Engineer and I were sitting on a hay bale watching the local kids playing this game. They were good. They were really good.

'Looks pretty easy,' I said jokingly to Scooter, who was whistling to his dogs as he strolled towards us.

'Put your money where your mouth is, Harri-Henry! Have a go.' Trev gave a roguish smile and handed me a child's lasso.

Well, what could go wrong? It all looked pretty simple. But despite my almost perfect swing, and the good feeling that the rope was going to easily flip over the metal head, I missed that calf by a mile. I tried again and again, and when I finally gave up, there was a spontaneous burst of applause.

'Stone the crows, there's room for improvement. Better stick to ya knitting,' Trev bellowed as the kids burst out laughing. They enjoyed the moment enormously. And for days afterwards, I was subjected to offers of loser's drinks and one-on-one tutoring, usually accompanied by knowing winks.

What's Important on a Cattle-Healthy Farm

Owning cows was a little intimidating at first. There seemed a lot to learn as we managed their general day-to-day care, adding lots of hands-on experience to our understanding of managing healthy calves and contented cows.

A cow is pregnant for roughly nine months. Our experience of cows delivering calves has all been rather easy. We just stroll out to the

Chapter Nine: The Easy-Peasy Method of Caring for Cows

cattle yards and there is a new baby. I haven't needed to worry about birthing issues at all, and the excitement of finding a strong new calf feeding from a healthy-looking cow, is really something special.

But my farming friends have told me lots of stories, so I sort of get the picture. I understand the importance of knowing how long the cow has been in labour, so a decision can be made as to whether to give the cow more time or call the vet to help correct a problem. And, for example, if there is no room to force a couple of fingers between the calf's forehead and the cow's pelvis, the calf is not going to fit, so it's definitely time to call the vet.

In early labour the cow is restless, getting up and down and kicking at her belly with her tail held out. Breaking of the water signals the beginning of active labour, as the calf starts into the birth canal and abdominal straining begins.

'You can't intervene too soon,' Lockman warned me. 'If the cervix is not fully dilated and you can only put one or two fingers through, the cow needs more time. So, wait until the cervix is dilated or you may injure the cow by pulling the calf through that narrow opening.'

He must have read my questioning look because he began to explain further.

'Put your hand into the birth canal as far as needed to find the calf and you may discover the feet are there. But then again, don't pull too hard too early. Forceful pulling before the birth canal is ready, may rupture the cervix or tear the vagina and vulva. The cervix opens as the calf's head presses intermittently on it with each contraction.'

I listened without much enthusiasm as Lockman continued. 'But … if you wait too long to help, the calf will die. Give assistance to the cow if she's been in early labour more than six to eight hours. Or she is straining hard for more than one hour, with nothing showing. Or if the feet show when she strains but then go back in. Or the feet of the calf look upside down, or if only one foot appears. If the calf's tongue is sticking out, labour has probably been too long, especially if the tongue is starting to swell. The cow may be exhausted by then and unable to strain productively when you do try to help her.'

As Lockman was bouncing these details around, Scooter turned up and joined in the conversation. I marvelled at their enthusiasm as they shared their preferred methods for 'pulling a calf' with chains attached to the calf's legs, using a half-hitch – one loop above the fetlock joint and the other around the pastern above the hoof.

'This spreads the pressure better than a single loop and will cause less injury. Then you pull when the cow strains and rest while she rests.' Lockman chuckled at my disapproving face.

Then Scooter said he always liked to have a helper stretch the vulva as he pulls, making it easier for the head to pass through. I was somewhat appalled by this, but he hadn't finished and briefly touched on what to do if the calf is coming backwards, giving me even more to ponder for one day.

'The plan here would be to attach chains to the hind legs – double half-hitch – and pull slowly and gradually until the hips are coming through the vulva, then to pull the calf on out as swiftly as possible so it won't suffocate.' Scooter cleared his throat and grinned reassuringly.

'Okay!' I was certainly not enthused by any of this and I looked at them with an expression that said, 'Yep. Thanks. Got the general understanding!'

At the end of it all, I decided that it was quite stupid to worry about what I would do if I was faced with the unnerving sight of a cow in delivery distress. I would be loath to make any assessment and consequently, if I had even the slightest inkling that there may be a problem, I'd be on the phone to Lockman or Scooter or Meggs.

In the winter months and during a drought, our paddocks lose condition as the grass turns brown and brittle, and we begin to hand-feed generous serves of good grassy hay, prime lucerne and sorghum or barley hay. The cows love their hay and when they see us coming around the side of the barn, they will tear up the hill, their funny short legs going flat-out, to line up at the gate and bellow out to us.

But on one Christmas Eve afternoon, Farah, one of our pregnant cows, was missing at her dinner time. I wasn't really worried, but almost straight away we set off to search for her. We looked everywhere and eventually found her in the hollow of a dried-out dam. She had just given birth to a lovely healthy calf, Baby Christmas.

After we'd finished admiring the newborn calf, it struck me that this was not a good place for the cow and calf to be. It had been raining all day and the weather forecast was suggesting the possibility of a severe evening storm. During the afternoon the rains hadn't presented any great threat, but I was concerned about our new calf lying in the bottom of an almost empty dam, with a storm approaching. I would have felt more comfortable if we could have moved her to higher ground, but Christmas was going to be too heavy for *us* to carry and it was a bit too soon for her to walk any distance. The thought of driving the tractor down the wet slopes was an unattractive option.

As we returned to the farm house, the wind and rain began to gather with savage force and I did my best not to think how dreadful it would be if the dam began to fill with water – and drown our new calf. The Engineer stood out on the deck and, peering through the storm, tried to estimate the amount of rain falling. He figured it would take several days for the dam to fill to a dangerous level and I believed him. Especially as I really didn't feel up to going out in the dark and stumbling around looking for a black calf lying in puddles of water.

Next morning was fine. The storm had blown itself out and there were no clouds in the sky. The early sun was rising as we made our way down the hillside, and to our delight, we could see Farrah on top of the dam wall, feeding her calf. Christmas turned around as we approached and looked up at us with interest. There were spots of milk on her cute little face and we got close enough to stroke her soft satiny coat. In no time at all, we easily walked Farah and Baby Christmas up to a high ground paddock, close to the house, where I could keep a close eye on them.

Three-Day Sickness

You never know what is around the next corner and I was caught off guard one day when Derek mentioned that three-day sickness was in our area, and as a conversation starter, three-day sickness takes some beating. I hadn't heard of this condition and I didn't fully appreciate what it really is. I listened carefully as he told me about some typical signs: the cow will be down and depressed, have a short fever with shivering, as well as lameness and muscular stiffness. And worse still, I began to hear snatches of talk about three-day sickness as I chatted to various folk in the shopping centre and they all had different viewpoints and opinions.

Did I sort of know what it all meant? Not really. And so it seemed a good idea to spend time on Google, where I learnt that bovine ephemeral fever (BEF), commonly known as three-day sickness, is a disease caused by a virus. It is generally believed to be transmitted between cattle by flying insects such as mosquitoes, buffalo flies and sandflies. The spread of the disease depends on weather conditions and when storms are prevalent, they can contribute to the rise in cases by providing good conditions for breeding biting insects. If there have been successive years of dry seasons with little biting insect activity, cattle with no previous experience of three-day sickness will have no immunity. At its worst, there can be losses due to this illness: lowered fertility, maybe occasional abortions, and in general, a loss of prime condition.

And there was more: cows can suffer complications due to related issues. For instance, in their weakened state they may injure themselves falling down a slope, or even into a dam, and not have the strength to climb out. Our farm has many steep slopes and I certainly didn't like to think what I would have to do if this happened to one of our cows.

It just happened that a few days later, after my chat with Derek, I was passing right by the newsagent and came across Trev climbing into his roo-battered** ute, with the morning papers under his arm.

'What's up mate?' There was a twinkle in his eye and he gave me a thumbs-up when I strolled over to say hello. After some small talk, I asked him about the three-day sickness vaccine.

'How many cows in your herd, Harri-Henry?'

'Eight.'

'How well is your farm paying?' He started to grin as my face registered my perpetually hard-up expression.

Chapter Nine: The Easy-Peasy Method of Caring for Cows

'You shouldn't worry because cattle that are well looked after, have some shade cover and access to feed and water, will usually recover,' he assured me.

Later that same week, I came across one of our cows, Gracie, huddled miserably under a tree. I stopped and considered her despondent demeanour. It seemed Derek was right and three-day sickness was indeed in our area. Gracie didn't look well, lying down with her head bent forward, and I recalled Trev's bush advice that cattle that are well cared for, have some shade cover and access to feed and water, will recover after a few days.

Fortunately, Gracie had staggered under a leafy tree to rest and so shade was not an issue. We did offer water. She didn't drink it, but the other cows did. We watched her carefully and as much as possible, eliminated all stress. We didn't try to move her or shift her, and to my great relief, exactly three days later, Gracie got up and resumed her grazing life as if nothing had happened.

After Gracie's episode, our little Ernie, Big Ernie's son, also came down with this sickness. However, he was only down for two days before staggering to his feet and lumbering off after the girls.

About Calves

I was a little worried when I noticed Elvis, our little bull calf, miserably trudging along far behind the other cows. His progress was painstakingly slow and he just didn't look right. The Engineer helped me coax him up to the cattle yards and we looked him over. Elvis wasn't showing any obvious three-day sickness symptoms and he seemed free of ticks. A snake bite was ruled out straight away, as too much time had elapsed and Elvis was still alive.

I knew I needed to do something sensible and so decided to phone Meggs, and I was thankful she was able to examine Elvis before dark. She wasn't sure what was troubling him but suggested a course of antibiotics. Then she showed us how to restrain him and where to give him the daily injection – for a week. After this initial treatment, Elvis began to get better. A day or so later it seemed the blood tests could only establish what he did *not* have, but fortunately he fully recovered and grew into a large and rather good-natured, sturdy bull, always to be seen following faithfully behind his mum.

Raising bottle-fed calves is really quite straightforward and may be necessary for a variety of reasons. The calf might simply be a twin and the cow may only have enough milk for one baby. Or it might be a heifer's calf that isn't accepted by its mother – or a calf whose mother has died. With a newborn calf, bottle-feeding is simple because it will be looking for milk. Where possible, the first feeding should be colostrum – the perfect food. As with lambs, colostrum has a much higher fat content than regular milk and gives the calf energy for strength and warmth, as well as important antibodies to protect it from various diseases in the first weeks of life.

A rubber lamb teat on a bottle works well for a newborn calf. Most produce stores stock these long rubber teats that can fit over most bottles. But for the bigger, older calves the stiffer calf teats are the best. We make sure the hole in the teat is not too small because if the calf can't suck enough through it, it will become discouraged. But it can't be too large or the milk will flow too quickly and choke him.

Lockman showed us how to bucket-feed poddy-calves. It can take a little persuasion, wriggling our fingers in the bucket of milk before the calves realise that this is yummy dinner time. Now, whenever we have poddy-calves and they see us approaching with any sort of bucket, there is a mad dash with lots of hopeful mooing and hollering.

Flies, Ticks, Worms and Other Issues

It's a fact that a well-structured drenching program is a must in order to manage internal parasites such as roundworm, flukes and tapeworm; otherwise cows can lose condition, and calves, most susceptible to the effects of worms, will be poor in growth and, critically, severe infestations can cause death.

An anthelmintic drug (a wormer, de-wormer or drench) is given to cattle to rid them of parasites through the process of drenching or worming. There's a stack of knowledge about it on Google: pages and pages of technical stuff; options and choices; the difference between good and bad practices; and even '*to worm or not to worm*' debates! Most confusing.

Meggs recommended we use a faecal egg count as a sensible way of knowing whether parasites are resistant in our cows. In this process, a sample of fresh dung from the ground is delivered to a veterinarian for parasite analysis or samples can be sent direct to laboratories to have egg counts done. The results help decide which drench is to be the most effective. The products Ivermectin and Moxidectin work by paralysing worms and can kill external parasites such as lice, mites and larval skin forms involved in summer sores. Moxidectin can penetrate the intestinal wall and is probably the most effective. And Ivermectin has been around for about thirty years and has been a most reliable de-wormer, although there is concern that some parasites are developing resistance to it.

My reliance on informal advice, word-of-mouth recommendations and professional advice all helped me make a decision to worm our cows with user-friendly pour-on treatments. Once the cows are yarded and led into the race, it is relatively simple to follow the instructions on the label and, using the recommended applicator, we pour the chemical in a long strip down the middle of the cow's backbone. Cattle should be weighed prior to treatment, to ensure the correct dosage is determined. We didn't have the equipment we needed to weigh our cows and my methodical flipping through some web pages proclaimed my ignorance until, thankfully, I found a useful weight chart for Belted Galloway cows, calves and bulls.

While internal parasites can be a problem after wet weather, smart grazing and pasture management can also help control the problem, with well-fed animals rotated between paddocks with a low stock density.

Vaccinations

Vaccinations play an important part in disease prevention. We vaccinate our cows when we vaccinate the sheep, using the same vaccine, which means that we can keep our costs down. The key vaccines used in Queensland herds are delivered as '5-in-1' and cover five diseases with appalling names such as pulpy kidney, black disease, tetanus, blackleg and malignant oedema.

A '7-in-1' vaccine covers the same diseases, plus leptospira harjo and leptospira pomona.

And there is another issue too. Tick fever. Cattle in cattle-tick-infested areas are at risk of developing tick fever, whether they are born and raised within the tick area or introduced in from tick-free areas, and this is a problem. But there are some reliable and practical tick-fever vaccines available for cattle. Some farmers refer to this process as 'blooding' – and these are ordered through a vet or agricultural department.

External Parasites

I was warned that external parasites need to be properly managed to prevent them impacting animal health and performance. I was surprised to learn that buffalo flies can irritate cattle to the point that they will actually stop eating, resulting in reduced growth. And although a low level of buffalo-fly infestation is tolerable, cows not treated for flies can get very sick and, on occasion, die.

Chapter Nine: The Easy-Peasy Method of Caring for Cows

I went to Google to research treatments for waging war on external parasites and found there are many options: back rubbers, dusting appliances, pour-on products, flytraps, and insecticidal ear tags and injections, to name a few.

1. Insecticidal ear tags

Insecticidal ear tags slowly release insecticides from the tags onto the shoulders, back and flanks of treated animals and are the most widely used method of buffalo-fly control in Australia.

The Engineer and I hadn't a clue about how to put ear tags in cows' ears and soon discovered it is much easier said than done. In fact, we were completely useless, despite our best efforts. The tags are labour intensive to apply and nobody warned us that they can be really difficult to get into the ear of the cow, especially when the cow won't co-operate.

We started with Farrah. She glared at me angrily as she shook her head furiously and snorted from deep in her chest. So, we then tried with Gracie, and our persistent and pathetic attempts made her very cantankerous. It became obvious that we desperately needed some help and I knew we would be well advised to phone Lockman.

I couldn't believe how simple it seemed when Lockman did it. The cows stood still and glared at us as his big powerful hands effortlessly whacked the tags in the right spot every time. When all the cows were sporting a brand-new ear tag, Lockman looked at me quizzically and said, with a real twinkle in his eye, that I shouldn't hesitate to call him anytime, if anything else happened – well – because this was work for real farmers!

2. Insecticidal injections

But there is more: insecticidal injections. Stumpy said that insecticidal injections should be administered exactly at the base of the ear and nowhere else. It sounded challenging. But Wally the shearer said he used insecticidal injections on his cows and, 'just shoved it anywhere'. But … the biggest problem according to Lockman is there is no room for error. He casually informed me that we would need to be very careful, because if we got it wrong, the cow would die. His words were profoundly discouraging – and we didn't even consider insecticidal injections after our feeble attempts with ear tags.

3. Back rubs

Back rubs consist of absorbent material soaked in a mixture of insecticide and oil, fed from a tank of sorts, which cattle rub up against. Their success as a treatment depends on the frequency with which they are used by the animals. There's no control over a dose per animal, and the farmer needs to check regularly to ensure that the chemical and oil mix does not run out. They are only economical on sites where cattle congregate, such as watering holes or feed areas.

4. Buffalo-fly traps

The Engineer rolled his eyes when I mentioned buffalo-fly brush traps, but I have to admit I was fascinated by this option. The trap is a clear plastic tent which makes a short, darkened tunnel through which cattle pass, and due to changes in the light level on entering the tunnel, flies move off the cows to be caught in cages attached to the tunnel sides. It is economically worthwhile over a five-year lifespan, for herds greater than fifty head. Not so effective for a herd of eight.

5. Backline sprays

This is the application of ready-to-use chemicals applied as sprays to the backline of the animal or to the full body. Although the chemicals are relatively cheap, multiple treatments are required throughout the season.

6. Plunge dips

Even the old-timers agree that the good-old-days plunge dips for fly control are labour intensive and too expensive to maintain.

7. Pour-on products

Because Insecticidal ear tags need to be regularly changed, I began

Chapter Nine: The Easy-Peasy Method of Caring for Cows

looking for a user-friendly treatment; something that was easy for retirees like us to manage. I asked around and pretty soon we were directed to pour-on treatments. Such a sensible solution. What I can tell you is that this is our preferred kick-ass method and pretty much our go-to plan each and every time. This *is* something we can use with eight cows. Easy to use, these pour-on products contain chemicals to help control buffalo fly, worms, ticks and lice, and they are available over any counter in any produce store.

*A crush is a metal alleyway that a cow walks into: a frame closes to trap its head and a gate closes behind it so the cow is held in position.

**Roo-battered: damaged from the impact of hitting a kangaroo on the road while driving

TEN

True Tales from the World of Living with Goats

I have long had a fascination with goats. Goats are like chocolate: they're addictive and no matter how many you have; you always want more. Our goats turned up on the farm, peered over the fence and seemed to approve of their new surroundings of green hilly fields, lovely views and funny, friendly sheep.

Love them or hate them, goats are remarkably practical. We've saved heaps on chemicals and labour as the goats have sculpted and transformed our hillsides by munching their way through formerly weedy acres. It's like a long-term, ever-evolving art project. They will browse on most weeds that sprout up, even the noxious Queensland fireweed. Except, like Tigger in the story of *Winnie the Pooh*, they're not fussed about thistles.

Because we knew that goats can get over, around, through and stuck in most kinds of fencing, The Engineer rolled up his sleeves and set out to find some fencing contractors capable of building strong woven-wire fencing around the goat and sheep paddocks. And also, to build a protective electric fence around the sheep and goat yards, barn and enclosures for protection from feral dogs.

Chapter Ten: True Tales from the World of Living with Goats

How Not to Have a Billy Goat

The most embarrassing episode happened one day when a cloud of dust whooshing up the road announced the hassled arrival of our near neighbour, Shorty. It's a good name for a big farmer and Shorty has the reputation of a pit bull where animal welfare is concerned.

I was a bit surprised to see his ute pull up alongside our fence. He wound down the window and yelled, 'Gidday mate. Are you guys okay? Is anyone hurt?'

'Hey Shorty.' I grinned as I strolled over to the fence. I like him a lot and at that moment I was looking for any excuse to have a chat after finishing some hard jobs around the farm. Never one for aimless conversation, Shorty's open and friendly face was looking bothered. He nodded briefly and began to question me about the screams and groans he said he was hearing, coming from our bottom paddock. He was sure someone on our farm was in serious trouble.

'It sounds bad, mate. We can hear someone screaming for help!'

'Really?' I looked at him in astonishment.

'I can't hear it now though. Are you quite sure everyone's all right?' He was most insistent. 'We can hear it from inside our house.'

Listening in vain for anyone yelling, I simply had no idea what he was on about. I looked at him. Then I looked at my cell phone and rang The Engineer to see if he was okay and explained the situation. But it seemed The Engineer was fine. So, it all seemed very strange.

Shorty looked at me a bit exasperated, almost disappointed, as if he was expecting a more helpful answer than 'no – we're all fine'. And to hammer the point home, he told me again about the cries and groans he had heard coming, he thought, from our paddocks, which sounded terrible.

I suggested he pop over the road to check out the property belonging to Joan and Jeffrey. They live just a few fields away and run a herd of Dexter cows. As I stood and stared after him, a thought began to bug me, and I began to cobble together some ideas – then I got it. I mean – I really got it. The truth of this embarrassingly awkward happening struck me with daunting effect, and it was with some trepidation that I watched Shorty chatting to Joan and Jeffrey.

They were obviously surprised to see him and I heard an outburst of laughter from Jeffrey. And, when Shorty finally drove off, the look on his face was unforgettable. I felt hugely mortified. But Shorty didn't seem at all put out. He gave me a cheery wave, shouted, 'Hooroo!' and I viewed his thumbs-up signal with a sigh of relief.

Nothing more was said and, of course, I should have known straight away. All farms have their challenges and where there is livestock there is trouble. Particularly so in breeding seasons. Ginger Nuts, our newest billy goat, was big, red and white; loud; persistent; undignified; and annoying. And the real problem was that Ginger Nuts had reached the age where he was sophisticated and grown up, and nanny goats in season were extremely desirable.

Ginger Nuts was the source of the screams and groans coming from our bottom paddocks; I could certainly appreciate how these cries may have sounded like a person in distress! Because, before indulging in head butts and posturing, he was communicating his feelings of passion with deep, meaningful groans. Our tatty-arsed young goat sounded like a drowning elephant. His deep, excruciating groans were extra loud, extra long and extra disgusting. His constant interference and ceaseless nagging put the nannies in a bad mood and their eyes narrowed and flickered with contempt.

But on the flip-side, this behaviour only lasts a couple of days and that's definitely a good thing. However, in reality this was a sure reminder of the earthiness of farming animals.

The Engineer? He just found it very funny.

What to Do When Things Go Wrong

I never assumed that keeping goats would be trouble-free, but even I was caught unawares when one day I heard piercing screams coming from the goat yards. The volume of sound was incredible. Something was dreadfully wrong and I bolted over to the goat pens and gasped.

Ginger Nuts was trapped with his back leg threaded into the side mesh panel, his body hanging down sideways and his head stuck under the bottom railing. His limbs were trembling and his bulging eyes were fixed on me. It was a grotesque sight and I panicked as the realisation burst upon me with frightening clarity that this was probably the worst situation I had ever had to cope with – on my own. The Engineer was out and there was nobody home but me – city-born me, who had a dream of living on a farm. There was no one else around except Belle, her wide eyes staring, fascinated at the upside-down goat. So how on earth was I going to sort this mess out?

There had to be something I could do. It was a huge challenge and I felt absolutely hopeless as I leaned over the panel and tried desperately to free Ginger Nut's foot – but nothing happened. He began to thrash about wildly as I grasped the piece of mesh to try and pull it apart. However, in reality this was never going to work. The mesh wouldn't budge. I tried again. Nothing. No movement at all.

Panic-stricken, I flew into The Engineer's workshop and looked around helplessly at the shelves and cabinets overflowing with building and engineering stuff. My heart thudding, I didn't have a clue where to start. There were tools everywhere. On the benches. Hanging on the walls. Scattered over lathes and a milling machine. In fact, The Engineer, to my knowledge, has every tool known to man.

Chapter Ten: True Tales from the World of Living with Goats

In agonised uncertainty, I rifled through shelves and drawers hoping to find some sharpish cutting devices I might use. But the search was taking too long and clearly I was out of my depth. So, I grabbed a long metal crowbar, an old fret saw and some tinsnips hiding under some chisels, and rushed back to my patient.

This time, the billy was quiet for a brief spell. But I could see his angry eyes looking up at me. His hoof was clamped into the mesh and I had trouble locating the wire. I struggled with frenzied urgency and began to bang away with desperate endeavour. Cutting. Sawing. Smashing the wire with the crowbar – again and again. It was exhausting and challenging. To make matters worse, I wasn't good with tools at the best of times. But there was nothing else for it; I had to keep going.

There was a moment when I was convinced I was doomed to fail and, by now, I was tiring from the efforts of trying to free this stupid billy goat. Then, it seemed that maybe I was making some progress as a little indent began to show. I persevered and bent over the fence again, continuing with more erratic sawing. All the time the goat screamed and moaned and struggled. His strained eyes in his haggard face fixed on me in a frightened stare.

It wasn't hard to guess how the billy had managed to entangle his foot. I'd had time to figure it out. Nanny goats in season! That was what had started this reckless situation: Ginger Nuts trying to climb the fence to get to the girls.

I was quite out of breath when at long last a piece of mesh broke and I was able to break the first wire. This was half the job done, but I was going to have to cut a second strand to free his hoof. I began again to smash and saw away … until at last I did it! I was able to wiggle his foot out and set him free.

My relief was overwhelming. Ginger Nuts collapsed in a heap on the ground and I felt like doing the same. Then he stood up, wobbling and staggering, carefully holding up his wounded foot. It was some moments before he was able to hobble away and, with a stiff-legged gait, began to head out into the middle paddock. He actually seemed unruffled by the whole event.

'I think he'll be okay,' I muttered to Belle, who had found the whole show most exhilarating. The crisis was over and I made my way back to the house, poured a cup of tea and sank into an armchair. The sudden quiet was deafening and I could hear birds singing. It had been a ghastly episode and I couldn't help feeling that maybe I should have chosen a more traditional style of retirement. These past few moments had persuaded me that farming animals was totally overrated.

'Anything interesting happen in your day?' The Engineer enquired on his return to the farm that evening. I swallowed hard and said honestly that it had all gone rather splendidly – but I needed to learn about tools. Did he have any suggestions?

Baby Goats

Every year our nanny goats have cute, funny babies, and one time, Nanny Doreen gave birth to triplets! This was our first multiple birth and I was utterly entranced by these pocket-sized babies. It was late in the afternoon and I had climbed a little way down the hillside to where I could see Doreen had given birth under a tree. Once I got over my surprise, I squatted down to stroke their tiny, smooth, red and white coats before phoning The Engineer with the news. Everything was just lovely, until we started to bring Doreen and her three babies uphill to the goat yards before night fall.

The Engineer did have the foresight to bring a large plastic bucket to put the baby goats in, and the plan was for me to carry the bucket while he encouraged Doreen to follow on behind. At least that's what we were hoping for – but I have found in my brief experience of owning goats that nothing goes to plan.

We might have guessed that as far as Doreen was concerned, no sensible argument was going to stop her believing her babies were still at the bottom of the hill. Indeed, we hadn't gone far before Doreen was off down the slopes with The Engineer in hot pursuit, while I continued to stump my way up the hill, with three babies in the bucket. They were heavy and I had to change the bucket from hand to hand every few minutes as the kids began to wriggle and jiggle, until

Chapter Ten: True Tales from the World of Living with Goats

the bucket began to tip over sideways, making it almost impossible to carry.

Separated from their mother, the kids were a screaming handful. About halfway up, as I was plodding slowly, I decided it was all too difficult. I grabbed the biggest baby and plonked him down on the grass. Now I had two babies screaming in the bucket and one hunched and screaming on the ground. At the same time, The Engineer gave up on Doreen and left her to sort it out for herself. As he climbed back up the hill once more, he grabbed the sobbing baby squatting on the ground, tucked it under his arm, and opened the gate for me to wobble through with my bucket of two babies, still screaming eye-popping, high-pitched wails and squeals without any apparent need to draw breath. Really! How could such tiny babies make so much noise? I just wanted to throw up my arms and shout, 'Hey Doreen! Stuff it! Listen!! I've got your babies here in the bucket.'

I came close to despairing that we'd ever reunite the four of them and it was nearly dark when The Engineer went down the hill once more to find Doreen. Then, without any warning, she about-turned and bolted up the hill at a furious pace with The Engineer panting along in the rear. It was a huge relief when Doreen rushed into the barn and began to tuck into her dinner of hay. And with the sudden arrival of their mother, the triplets stopped bawling and began to squabble over her bulging teats full of milk. Later on, as they snuggled into their straw bale beds, they looked so peaceful and cute, and I told The Engineer that I thought it had all gone rather well.

The following day, Nanny Mavis produced twin girls. There was no fuss. No buckets. No screaming. And shortly afterwards, I watched Nanny Mavis, accompanied by two delightful babies, sashay into the barn with perhaps justifiable pride. Now this was nice. It was a much better moment.

The Story of Spotty and Dotty

I am never at my best when I have to get up in the night – in fact I am considerably grouchy. And I was sound asleep one filthy wet night when I was woken by the plaintive cries of a newborn baby goat. I looked at my phone. No kidding, it was 2.55 am. All the same, I knew that a lot of things can be going on under the cover of darkness, so I groggily stumbled out into the freezing cold, rumpled top to toe in outdoor pyjama wear, my pockets filled with necessary items such as a torch, gloves and paper towels.

I opened the gate and went into the barn to find there was plenty happening. Triplet goats had been born while we were sleeping and I could see one of the babies was already dead, leaving two – Spotty and his sister, Dotty. This was a really bad moment. Dotty was bright and alert and beginning to understand the drinking thing, but tiny, miserable Spotty lay limp, flabby and listless – not a good look. The appalling misty, cold weather didn't help and so I carried both babies inside to the warmth of our reverse-cycle air conditioner. As I cuddled tiny Spotty, bundled up in a thick warm towel with just a wee pink nose and two anxious wide eyes peeping out, I tried to get rid of the nagging doubts and thoughts that he mightn't make it.

Spotty's first few days were pretty horrible and at first I couldn't see a way of helping him. He was too weak to suckle from his mother

Chapter Ten: True Tales from the World of Living with Goats

so every few hours I bottle-fed him tiny amounts of colostrum that I milked from the nanny. He developed a tormented, hacking cough and repeatedly gasped and wheezed, and I tried not to feel despair as I massaged his tiny, heaving chest. Given these issues, it's hardly surprising my mind pondered on the possibility that the triplets would be reduced in number – to only one surviving baby goat.

I had no real expectation of finding him alive each morning when I went out to the goat pens. But despite the grim outlook, I continued to feed him and massage his tiny chest before lying him down on fresh straw in the warm winter sun. What worried me the most was that he just looked downright miserable.

A breakthrough came late one morning: it was a sunny day and Spotty looked awful as he lay in the warmth of his straw bale bed. Then to my delight, he struggled to his feet, tottered over to his mother and began to suckle. I was SO excited. It was clear that the crisis was over and there was no doubt he was enormously improved.

From then on it was a gradual process and, every few days, he seemed a lot better. He never quite got rid of his cough, even as an adult billy, and if the evenings were cool, we would often hear him gasping and choking and struggling for breath. Despite this, it was a spectacular result even if it seemed that he'd have his wheezing and snuffling for life.

Goat Management

For the first couple of weeks after delivering their babies, the goats hang about the pens and we notice that the kids suckle most of the time. Their eagerness for milk ensures them a steady supply and this contact between mum and babies, for the first six to seven weeks of age, helps to maintain a high milk production. But after we let them out into the grassy paddocks and the kids' intake of natural grasses builds up, we have seen the goats increasingly reject or interrupt suckling sessions, marking the start of a natural gradual weaning process.

We worm our goats when we worm the sheep – the dose is at 1.5 to 2 times the sheep dose (depending on the drug). And in spite of the goats browsing on stony hills, we do have to trim their feet to help prevent the hooves growing too long or becoming cracked and split. Trimming Big Billy's feet is a challenge and, more often than not, we have to tie him up first in the cattle yards. Ginger Nuts is usually easy to manage, but can be troublesome and stubborn when he doesn't get his own way. Sometimes he is more effort than the job is worth, so he wins. It is far easier to pass on this chore to Wally, the shearer!

The Particular Sadness of Delores

'There'll always be things happening with young goats: they're stupid, too daring, thrill-seeking, dumb animals.' With the utmost good humour, Aunty Tup didn't hold back on what she really thought.

I protested good-naturedly as we leaned over the gate to watch the young goats hoon around the hay barn. At the time, I was trying to point Delores out to Aunty Tup. Delores was my favourite, always full of comical tricks, as she bounced along like a kangaroo on a pogo stick. But her story is one of misadventure, courage and reckless self-confidence. It all began one day when I thought I could see a large white plastic bag caught in the fork of a scrubby tree, way down in the bottom paddock. And it was not until I took a closer look through binoculars that my stomach gave a lurch: Delores was dangling upside down, in the tree, trapped uncomfortably by her front foot.

I tried not to freak out as The Engineer drove the tractor down to the bottom paddock. He positioned the tractor's bucket underneath the bough of the tree and between us, we were able to get her free and put her in it, for a gentle ride up the hill. Back at the sheep yards, Delores looked miserable and shaken, with her foot hanging awkwardly at her side. She peered at us anxiously as The Engineer bent down and ran a gentle hand down her leg. As far as we could tell, she hadn't broken any bones and she was able to stand on three legs, but there was no doubt her foot was unable to support any weight. I looked at her and I didn't know what I could do.

The following day things were no better. But after a few weeks of just standing around, doing nothing, Delores showed an eagerness to

Chapter Ten: True Tales from the World of Living with Goats

go back to the herd, despite her foot issues. I had about given up hope but it was really a shame to keep her separate from the other goats, so I opened the gate. Delores looked around and headed purposefully, hopping and stumping, over to where the other goats were grazing in the paddock, and it seemed perhaps that we were back to normal again as Delores settled back into the goat herd.

It was nearly a year later when Delores, as cool as a cucumber, strolled out into the paddocks as if nothing had ever happened. You'd have thought we'd imagined the whole deal, because she had absolutely *no* sign of a limp and I must admit to having a sense of pride at her obvious return to normal movement.

As the drought worsened, we became stuck in the midst of one of the driest and hottest summers in our experience and, on an impulse, I decided to move the goats from the middle paddock into the next field with the cows. It is a benefit for goats and cows to share paddocks, because cattle eat the taller grasses while goats eat short grasses. Even so, there wasn't much grass left for the animals to eat and we were having to throw out hay each day for the cows. At first the goats appeared to be totally disinterested in the hay spread out over a wide area, and I did assume the animals would share!

I was wrong.

After exploring their new paddock, the goats headed towards the hay strewn across a gully and I saw Delores plunge determinedly into the midst of the feeding cows. I could hardly believe the speed with which one of the cows swung around, and with a mighty butt of her solid head, sent our little goat flying through the air to land with a thud in the dirt. Delores never really had a chance. There were no signs of injury but she was killed outright. It was horrible – one of the most heartbreaking things to happen on our farm.

Well, this wasn't covered on Google – I'd never read anything about *not* having goats and cows in together. No one had thought to tell me *this*! As I climbed the farmhouse steps, my despair was interrupted by the welcome appearance of Scooter. He was strolling up the driveway and I stumbled forward to meet him.

'I've had a disastrous start to my day.' It was obvious I was extremely upset.

His expression did not flicker as he looked at me thoughtfully. The first thing he said was, 'Tell me what you did.'

I tried to ignore the huge lump in my throat as I explained it as well as I could.

'Shit happens,' Scooter said. His voice was steady and conversational. 'You've moved away from your old comfortable, predictable and traditionally ordinary life to farming animals for the first time.'

I sighed and looked at him helplessly. 'I don't know what I'm doing anymore. It's never the same as it is in books. I don't know what I'm going to do.'

Scooter considered this in silence. Then he said, 'Don't think too much about it. Experiences like this will go to making you a better farmer. And now you know why cows have to be moved *out* before you organise goats or sheep to graze the cow paddock. When cattle and sheep or goats graze, or even drink together, cows can be pretty rough. I've seen sheep pushed into water troughs, and even worse – when ewes and lambs run with cows and calves, sometimes cows will pound the lambs into the ground. Next time you'll get it right. You've got all the things it takes to succeed – a love of the land and animals – and a commitment to keep working when other retirees are enjoying their leisure. And your secret weapon – Engineer Man!'

I stared at him and cringed inwardly. This was different; I couldn't dispel the memory of this gorgeous little goat with her healed foot, lying dead in the dirt. I wish I had known not to have cows and goats together in the same paddock.

Next morning, I reassessed Scooter's words. My dream of being a farmer was travelling over some bumpy bits, but I was slightly mollified by his thinking. And, even so, I still considered this the best retirement lifestyle to be signed up for.

Chapter Ten: True Tales from the World of Living with Goats

ELEVEN

Ponies? Come on!

It was surprising how quickly the first few years on the farm flew by. I had a warm sense of belonging as a real farmer sliding into place as I began to feel more confident about handling and managing our stock. At the same time, I did attempt to display some restraint in terms of collecting appealing animals. I did say no to adorable baby alpacas. And I did say no to a tiny donkey called Rufus. But there have been exceptions and I've done my fair share of spontaneous buys.

One of my most memorable purchases happened when I set out to buy some guinea fowl keets* and returned home with three irresistible miniature ponies and one five-month-old miniature colt. I had heard that miniature ponies were easy to look after and quite gentle by nature; exactly the excuse I needed, given that I was quite unable to resist these cute little horses.

I did consider the thought that The Engineer might not be best pleased. But even *I* was a little surprised at his reaction.

'Look what I found at Cherie's Chook Farm,' I said as I led the ponies out of the float. 'Can I keep them?'

There was a pause and then, like an arrow – straight to the point – he grumped, 'Are you !?@??! # kidding me?'

I couldn't think of an answer to this and, to put it mildly, I was disturbed. He is such an unflappable guy usually, but it seemed he didn't

Chapter Eleven: Ponies? Come on!

share my enthusiasm for ponies. That was until, a couple of days later, I snuck around the corner of the hay shed and caught him patting Molly. 'They are rather special,' he conceded with an unashamed grin. I have to admit that I turned around so he couldn't see my *'I told you so'* face. Then I sort of pretended that it was his idea for our cheeky, playful little horses – Molly, Dolly, Colin and Jeffrey – to stay.

Should I Have Seen Trouble Looming?

I knew nothing – nothing at all – about managing a real live five-month-old colt and soon I realised that this tiny and gorgeous honey-coloured foal called Colin was not cute and loveable, but a randy little bugger. He was feisty and unbelievably strong. I couldn't catch him or pat him: now I didn't even like him. My mind was filled with misgiving as I began to feel quite overwhelmed by his wild antics and it wasn't too long before I had huge regrets about bringing him home to the farm. I was at a loss to know what to do and this was quite laughable, considering that in the beginning it had seemed such a good idea. I never imagined that such a tiny foal could be so much trouble and a feeling of total uselessness gripped me. I had no hope of ever being able to manage him, I reflected gloomily, and my mind was now consumed with how I could get rid of him.

Well, it just so happened that as I was trying to pluck up some courage to admit my failings to The Engineer, our local handyman, Bill, popped in to show me his new mobile chook tractor. He stayed for a chat and I was absolutely flabbergasted when he spotted Colin and straight up asked if I would sell him.

'Really?' I blurted out. My face must have shown my astonishment as I leaned back against his van and shoved my trembling hands into my coat pockets.

Bill shrugged with a faint smile. 'He'll be perfect for the kids, mate.'

For a brief moment I hesitated. Then followed my mad scramble to convince Bill that if he took Colin away – there and then – which meant now and immediately – he could have him for nothing.

It's hard to describe just how effortlessly Bill cornered Colin in the yards, grabbed him, and plonked this baffled colt into the back of his rather small delivery van. It was incredible to watch. It's funny how things fall into place, I thought, as I waved off the little white van and its unusual cargo. I was SO thankful – although oddly, part of me felt a little sad.

I didn't hear anything about the foal for months. Then late one evening, Bill phoned to say how much they all loved the colt, how big he had grown, how the kids could do anything with him, and that they had renamed him Fergie. He invited me to come to the school markets the next Saturday to visit him. And I did too.

There was Bill with his fruit and vegetable stall, and Fergie tethered contentedly to the stall railing. Fergie was picture perfect with his shiny honey-brown coat, golden mane and delicate, tiny head as he munched happily on his bucket of goodies. I looked with pride at the beautifully groomed little horse, wearing his soft green halter and gazing at me with calm, warm, brown eyes. It was difficult to believe it was the same feisty foal.

Chapter Eleven: Ponies? Come on!

I felt very grateful to Bill for providing such a wonderfully suitable home for this pony and, much later, I learnt that Bill was teaching Fergie to pull a little cart for the sheer pleasure of giving the kids some fun, and also to carry their market garden produce.

And anyway, I told myself, we still had Molly, Dolly and Jeffrey to have fun with.

It happened that as we were learning about smart paddock management for goats and cows, we were also genuinely surprised to find out that it's not a good idea to have tiny ponies and sheep grazing in the same paddock, even if they seem to be much the same size. One day something awful came about: our sheep and lambs were grazing happily, then without any warning, Jeffrey was off at full gallop, charging aggressively into the lambs. It was such a chaotic scene.

The Engineer said '?!%!*#!' and ran out into the paddock in pursuit of Jeffrey. Appalled, I pulled on my boots and kind of whooshed down the hill after him. Hair tousled, eyes glazed, caps askew, we flung ourselves into automatic action, our retired old legs flying across the slopes and gullies like elite athletes. Just in time, we managed to rescue some twin lambs at risk and after this experience, I was so disappointed with everything that seemed to go wrong that I fell on the couch, sipping a glass of something good.

I thought it must be obvious by now that I really knew nothing about farming animals. So, next day I talked a little with Ross. I couldn't help liking Ross. He's tall, grey-haired, and across his left forearm is a tattoo of a bull rider flying through the air with a bucking horse bouncing over his head. Always happy to help with any sticking points, Ross knows horses. What works, what doesn't. We said our hellos and then my voice took on a great weariness as I told him the whole story.

'On a farm there's a number of compelling reasons to make the right decision and you take a risk when you run horses with sheep.' Ross raised his eyebrows and half chuckled. 'It would be nice to think they can be best friends, but a similar thing happened one day at home with Kathleen, our sweet-tempered old mare chilling in the backyard close to the house so the kids could ride her after school. Well, Kathleen took a whim one day and chased some young ewes and fat lambs into the dam. It was difficult to believe this sweet pony could look so formidable. And the commotion that followed! From where I was inside the workshop, I had no idea what was going on – until I heard the frantic yells from the kids.

'"*Dad! Dad! Daaad!*" Everyone was shouting at once. The kids were screaming. The dogs were barking. The wife, Annie, had a full-on view of everything from the house and she was yelling.

'I jumped on the quad bike and roared off with the dogs down the paddocks. We didn't have far to go and the kids dived into the dam and scrabbled around to rescue the lambs. We all got soaking wet shooing the horse away, as the bellowing sheep clambered up the bank behind us. The kids did manage to get the lambs back to the house. But then there was a huge row because they used *all* the bathroom towels drying them out. The wife wasn't happy. I was in the dog house. We never trusted horses and sheep together after that.'

I nodded in silence. This was the kind of farm story that was absolutely thought-provoking, hilarious and a powerful lesson in farm management.

*keets: guinea fowl chicks

TWELVE

How to Keep Guinea Fowl – and Why

As I looked at Chloe, our German Shepherd, reeling across the grass, I instinctively knew that she had been bitten by a snake and a mixture of horror and disbelief surged through me. We made the vet surgery just in time and a week later Chloe was discharged, having made a slow but full recovery. But now everything was different and I began to worry about snakes lurking around in the gardens and under bushes, just waiting for an opportunity to get me and my dogs.

Aunty Tup came to the rescue. 'Get some guinea fowl!' she told me reassuringly. And she began to tell me about the large brown snake that her nephew, Gary, had unexpectedly seen heading for his stables.

'He's got Arab horses!' Aunty Tup said. 'And he's mindful of the value of guinea fowl around his stables. Gary said a "big brown" was heading towards his precious foals one morning and he watched an angry group of guinea fowl attacking it. They killed the bastard! He told me later that it was like watching a footy team in a scrum: something on a spectacular scale.' Aunty Tup paused. 'I'd like to have seen that.'

Aunty Tup's story started my interest in guinea fowl. They originated from Africa and still exist in the wild, but for hundreds of years they have been kept domestically in many countries throughout the world.

These funny, charming and quirky little creatures are one of the most self-reliant poultry birds around, and as well as being a definite deterrent for snakes, they are one of the best natural solutions for paralysis ticks.

I did know that paralysis ticks are deadly, but then again, it wasn't until we had a close encounter with a paralysis tick that I knew I absolutely *really* had to have guinea fowl.

Our tick story began one afternoon, when I caught sight of a tiny black twin lamb wobbling and staggering across the paddock. I felt some alarm. It shouldn't have been hard to guess what the problem was, but I didn't associate the stumbling gait with tick poisoning. I picked up the lamb and it did seem poorly. I decided to take it in to the vet to see why it was going downhill so quickly. And my decision was a good one.

I handed the lamb over to Trish, the vet nurse, and straight away she saw the problem.

'Hey! It's got a tick!' she exclaimed. Her eyes behind her thick spectacles were wide with concern as she ushered me towards the consultation room where we weighed up some options and agreed on a really low-budget treatment.

It was such a simple explanation for the lamb's ill-health – a paralysis tick – and now I could see it too: a small dark spot hanging off the lamb's jaw.

'And finding a black tick on a black lamb is harder to find than a dam full of water in a drought.' Scooter fell about laughing. He'd popped over later that evening for a cuppa and I knew he would never *ever* have taken a baby lamb to a vet!

Next morning when I collected the lamb, Trish said laughingly that she had decided to call her 'Baa Baa Black Sheep'… after the kids' nursery rhyme. And the receipt for payment listed the client as 'Baa Baa Black Sheep'!

It did seem a good idea to take home an overflowing carton of tick collars – one for every lamb to wear for the remainder of that summer. But as our flock of lambs grew in number every year, tick collars were only going to present an expensive and very temporary

Chapter Twelve: How to Keep Guinea Fowl – and Why

Band-aid solution. I was conscious of an increasing concern. I needed to get some guinea fowl as soon as possible.

I told Aunty Tup that I would feel a lot safer with guinea fowl around, grazing in the paddocks all day and eating entire armies of bugs and unwanted insects.

'Course you would. I can tell you that!'

So, I asked around and someone remembered that Fred Bullock was the go-to bloke for guinea fowl keets, and carried along with enthusiasm, I made a few phone calls and organised Charlene to accompany me on a visit to Fred's property.

His place was not easy to find and it was a long rough road. On the way, Charlene phoned Fred and said that if he could see a white van cruising at a snail's pace up and down the road, it would probably be us. And Fred climbed into his old car and came to look for us. His driveway was a bit concealed by a sharp bend down by the river, which was our excuse for driving by it a couple of times.

He had a rough wooden sign nailed on his old gate which in faded black letters spelled out *'Guineafoul for sale'*. And the number 102 was scribbled on a rusty milk churn.

'He can't spell then,' Charlene said. I was getting used to tatty farm signs on farm gates, but I did screw up my face when I read this one.

We drove slowly behind Fred, down a long ridged and rutted dirt track that weaved across the paddocks and through scrubby bushland. Soon we could see the farmhouse, just visible under a canopy of trees – a dilapidated worker's cottage where Fred lives with his brother, Reg. Neither of the blokes are married and though quite different personalities, get on well enough to enjoy breeding silkie chooks, turkeys and guinea fowl. Old Reg is nearly ninety but not really showing the slow stiff movements of an old man and he is known for rarely wearing boots. Even in winter, flannel shirts, baggy grey trousers held up very high by blue hay-twine braces, an old ragged cap and bare feet are Reggie's preferred farm wear.

We parked the van in front of a wooden barn beside the cottage, and as we got out, Fred motioned to us to follow him. Inside the barn it was quite dark and it was a few minutes before our eyes adjusted and we could see an assortment of mixed coloured keets and lots of silkie hens pottering around. A row of wire netting runs was wedged in between stacks of brown grassy hay bales and piles of clutter.

Fred leaned heavily against the hay bales and looked at me with raised eyebrows. 'Righto! You're on your own with this. Can ya manage?' He grunted and gave me a sideways glance as he made a non-committal gesture towards the rear of the barn and, with a sinking heart, I understood that if I wanted guinea fowl keets, I was going to have to catch them myself.

Charlene, clutching our cat cage, giggled as Fred, settling himself down into a comfortable position, began making somewhat helpful comments, which I couldn't really hear as I began to half crawl towards the chook runs. And Charlene, crinkling her nose, began to sneeze as she crawled in behind me, dragging the cage. She has such an infectious laugh and I could hear her giggling away as she called, 'Chooky-chooky-chooky, chooky-chooky.'

Chapter Twelve: How to Keep Guinea Fowl – and Why

It wasn't easy. The squeaking keets began to run away and hide under rotting timbers at the far end of the cages, and this was not the moment for me to wonder if there might be snakes hiding inside the piles of old firewood.

The keets put up a lot of wild panic-stricken resistance when I edged closer, my gloved hands outstretched ready to pounce, and not surprisingly, the free-range silkie hens squawked past in protest. But after much scrabbling around, to my astonishment, I caught one keet – and then another one – and clutching these peeping keets to my bosom, I thrust them at Charlene, who then flung them with surprising gentleness into the cat cage. From her crouching position, she effectively blocked the ones that escaped my grabbing hands and funnelled them into the corner, which was good because then they were quite easy to catch.

The Engineer had suggested a strict quota, which I ignored spectacularly as we crammed thirty-seven keets into the cage. As we stumbled out of the barn, the bright sun was already intense. Fred nodded patiently and stared at us as we struggled past, 'Good on ya, girls!' he said with a cheerful nod.

By now, it was a scorching hot day and the sky was filled with threatening clouds as we began our long journey home. Such was the beginning of my close relationship with guinea fowl.

Life became full of unexpected surprises after the arrival of our first guinea fowl. One most extraordinary experience unfolded one day, when I spotted a mother wood duck desperately steering her frightened babies – nine tiny wood ducklings with little legs pumping furiously – across our goat paddock to the safety of the dam and a screen of grasses and rushes. Flying almost on top of the ducklings was a kookaburra, unwinkingly eyeing off a potential breakfast. This was going to be trouble and I couldn't do a thing about it. At that very moment, the guinea fowl began zooming down the hill and I stared at the scene with something akin to awe. It was hard to believe how rapidly the guinea fowl were able to cover the ground, reaching the ducklings and shielding them with the intensity of their presence. They kept the kookaburra at bay with an outburst of ferocity, until the little wood duck family reached the comparative safety of the dam.

Then Derek, our postie, stopped to chat one day and we got to talking about guinea fowl – as you do when snakes and ticks become the main topic of conversation. Derek had recently enjoyed an African adventure holiday and he described one of the more memorable sights he had seen. Standing on a jeep during a wild animal tour, he had watched a sizeable collection of guinea fowl aggressively hustling five leopards out of their bush hideout and moving them on – at speed – across the grassy plain.

I enjoyed Derek's story and told him about the fox we had seen creeping up on one of our smaller guinea fowl hens, down in the laneway. A group of our guinea fowl, screeching and protesting, had circled the fox and whenever he lunged at one bird, other guinea fowl attacked his unprotected back, until eventually the fox got the message. He tucked his tail down and slunk away through another paddock and across the road.

Even a small, overwhelmed wallaby, just trying to mind its own business near our boundary fence, was hustled off the property.

To guinea fowl, wedge-tailed eagles are suspect. The eagles can appear out of nowhere to float low over our home paddocks. Their flight feathers stretch out like giant fingers and cast shadows over the gathering assembly of disapproving guinea fowl. The warning screeches of the guinea fowl go on and on as they glare up at the sky and, I must admit, this really shatters any peace and tranquillity. But it is usually short-lived as the eagles make a few reconnaissance fly-overs before gliding up and away.

Despite their courage and determination, guinea fowl living a free life are subject to attack from foxes, mysterious disappearances, or death by cars as they stroll across our dirt country road. These silly creatures have an absolute belief that all cars will slow down and drive around them.

Needless to say, on occasions we sometimes need to replenish our flock with new baby keets. Once they get used to their new surroundings, guinea fowl keets settle in well. To help them become familiar with their new home environment, we pen them up until they are a reasonable size. We don't usually take on mature guinea fowl, as they need to be penned up longer to prevent the urge to fly away at the first opportunity.

Guinea fowl will start laying when they're about a year old, and they lay for a few months during spring and summer. They typically find a good hiding place that is shared by several hens, so by the time I actually find a nest it's likely to contain lots of eggs. I found that if I removed all the eggs, the hens would find a new hiding place.

They are not great parents. They will lay their eggs in stupid places, anywhere and everywhere, throughout the spring and summer – especially in the long grass where it is difficult to find the nests. We did attempt to incubate eggs, using a small egg incubator we purchased from the produce store. Disappointingly, our success rate was poor. Very few of the eggs actually hatched a live chick and the chicks we did manage to hatch, never seemed as strong and healthy as naturally born ones.

If they do survive foxes, the tiny long-legged naturally born keets will trail behind their guinea fowl parents all over the paddocks, where they are vulnerable to being greedily gobbled up by ever-vigilant magpies, kookaburras and crows. Unless I intervene. If I can catch these speedy baby keets, I like to raise them under a heat lamp. At least this gives them a chance of survival. Our heat lamp is a fairly warm incandescent light bulb placed inside a large wooden box with a covering lid. We can adjust the heat by shifting the lid, so we can keep them warm on cold nights and reduce the heat on the warmer days. It's important to provide keets with a non-slippery surface so they don't slide and injure their tiny legs. Without sturdy legs, the babies will not make it to adulthood.

As the keets grow and develop, we move them into various sized pens until they are ready to free range. Even then, they need a place to roost at night. Our birds prefer to fly up to the top of our tallest trees to perch in the branches. If we stroll out at night and shine a torch up into the high canopy, we will see lots of plump bodies, funny heads and bright eyes glaring down at us. And in the darkness, we can often hear soft chuckles and faint rustlings and fluttering of wings. They know our voices and so don't make a sound, unless we bring someone new along to view them in their night habitat. Then word gets around and they commence their loud cries of warning and danger.

I've noticed that there seems to be an ordered system for roosting. It seems that the dominant birds fly up to roost first, and the ones of lesser status are last and assigned to perch on the lower branches. And, I've observed a similar pattern for departure in the mornings. It seems the more important birds at the very top of the trees fly out first, followed by the others from the lower branches.

One night, our guinea fowl began screeching a right ruckus to the point where I felt I really did have to get out of bed to see what was happening. I grabbed a torch and went outside to have a look, but couldn't see anything to cause the birds so much concern. However, in the early morning light next day, I could see three young guinea fowl dead on the ground. They had been chewed somewhat and it seemed likely a fox had been in the right place at the right time. These younger birds, I'm guessing, had stayed down on the really low branches and had paid a rotten price for their poor choice. I do try to keep young ones locked up at night for as long as I can – but eventually they have to learn to live with the older, more mature birds.

I had often wondered how guinea fowl managed to care for their young chicks, when in my experience they have not shown themselves to be great parents.

Then one day my friend Ginny rang. 'I've got a good story for you,' she chuckled. It seemed that Ginny had forgotten to shut the chook yard door and her one and only remaining guinea fowl hen, with three tiny long-legged chicks, had wandered out into the garden.

'I was so worried,' Ginny said. 'I watched them scurry across the paddocks into some long grass and all I could think of was foxes! And kookaburras! And how tiny and vulnerable the chicks are. But I couldn't have stopped them, even if I'd tried. I was beginning to think that they'd gone for good. Then I could hardly believe my luck. Next day, they all came back. But then I worried about how I was going to

Chapter Twelve: How to Keep Guinea Fowl – and Why

catch them before dark. And then, in the middle of the afternoon, I stopped worrying. It didn't matter anymore, because I looked upwards into the trees around the house and saw the guinea fowl hen showing her babies how to climb up an old, gnarled, ironbark trunk. The rough bark made little footholds for the keets to cling on to as they followed the mother hen to the very top. I'd never seen anything like it. And every afternoon after this, they climbed into the same tree and roosted in the security of the leafy branches. It was awesome and I couldn't help wondering why I'd fussed so much!'

Although adult guinea fowl are self-sufficient foragers, I do put out chook food to encourage them to hang around the house. Turkey starter is our preference for babies and when the keets begin to grow adult feathers, I add adult chook food to the mix. Although they are not tame, guinea fowl recognise our family members, including the dogs and cats, and will rush up in the morning to greet the daily arrival of grain placed under their covered container. And for worming we add a solution to the drinking water as per the product instructions. Chook worming products are easily found in our local produce store.

Our guinea fowl love to stroll around the house, pausing from time to time, to tap loudly on our tinted windows to admire their reflections. The first time we heard the tapping, we were most amused – it is almost as if they are inviting their reflections out to play.

I unashamedly love guinea fowl. These little guys are comical – small, light, whizzy and ever so useful. We feel secure in the knowledge that now we are 'sort of' safe from snakes and ticks.

'So, what do you think about having guinea fowl around your farm? A good idea – or what?' Aunty Tup nodded knowingly as we met outside the café one evening.

'Well yes,' I smiled. 'I've got to admit it – you're so right.'

Our small herd of Belted Galloway cows, including Big Ernie, the bull

Our sheepdog, Belle, bringing the cows up to the cattle yards

Belle herding Elvis, our young bull calf

Our sheepdog, Belle

A cattle crush is essential in the practical management of cattle

Side view of the cattle crush with the newly installed 'vet gate'

Belle playing after work with Charlie, our German Shepherd

Belle moving the mob from one paddock to another

Early morning muster on top of the hill

Our Boer Goats posing for their photo

Doreen and her triplets

Our guinea fowl enjoy a quiet moment

The guinea fowl eating ticks and checking for snakes out in the paddocks

The fascinating beauty of our hilly farm

A beautiful evening sunset

A scary summer storm rolling in

THIRTEEN

When Mr Bond Blew Away!

Our two cats, Mr Bond and Miss Moneypenny, live a free, natural life but their first duty is to rid the barn and hay sheds of rats and mice. I do appreciate the fact that in between sleeping on the couch all day and eating, our cats still manage to find enough energy to catch rodents.

One squally afternoon, a violent thunder and lightning storm rolled across our valley. It came up quickly, with great grey banks of clouds hiding the range of distant mountains. The wind was roaring through the trees and I was surprised how noisy it was outside. It was such a windy day, I could hardly stand up, and with the cold wind blowing hard against my face I was pleased to dash inside. I could hear the wind buffeting the old wooden farmhouse as I crouched in front of the fire to get warm. However, my curiosity was sparked when I peered through the window and caught sight of what seemed to be a large white sack being buffeted about in the wind. To my horror, I realised that this white blob being tossed high in the air was poor Mr Bond.

There didn't seem anything much I could do until the ferocity of the storm passed over and I watched helplessly, as our little white cat literally blew away down the hillside, to be dumped unceremoniously in the bottom paddock. I dragged my eyes unwillingly from the window, grabbed a coat, plonked a cap on my head ... and waited for the storm to ease.

HARRI-HENRY'S FARM

Later, as I trudged out from the house, the wind caught at me and I had to bend over almost double as I began the desperate quest to find our little cat. Squinting through eyes nearly closed against the driving rain, I was drenched and cold and longing to be back safe inside the house, when over the roar of the wind I heard Mr Bond meowing. I nearly fell on top of him as I combed wildly through the long grass. He was almost unrecognisable. Crouched in a ball, half protected by a fringe of trees, his face was muddy, his pupils dilated, and his expression was one of horror. I stared down at him, pop-eyed, then gathered him up in my arms. It was wonderful to get back inside the house, the wood fire was glowing and the lounge was warm and cosy. 'What a storm!' I said to The Engineer, as I handed him the cat to rub down with a warm towel. I collapsed wearily on the couch to pour out my story of Mr Bond's flying trip of terror.

That evening, Mr Bond curled up in front of the fire and didn't seem too distressed at all, although most of the next day he stayed hiding under the bed. Anyway, I was happy to think the worst was over.

I was wrong.

It was quite some time later when Mr Bond stopped eating and lay stretched out on top of the ironing board. The little fellow, with his

Chapter Thirteen: When Mr Bond Blew Away!

eyes closed, was so miserable and too sick to respond to our soothing words and attempts at stroking his head. I became seriously worried and decided I should take him to Meggs. She lifted him gently onto the surgery table and after a thorough examination, my fears were justified. This was not going to be a low-cost visit.

The diagnosis? Mr Bond had developed a urinary tract complication and the typical underlying cause of this problem – is stress!

'Really?' I said without enthusiasm. Meggs asked me if our cat had been under any stress at any time and I sighed. I felt a rising misery at the mental picture I was seeing as I described Mr Bond blowing away in the wind. Now that would've been stressful.

We discussed options and fees, and I stood for a few moments deep in thought before agreeing on the recommended treatment. The operation, a relatively straightforward procedure, would fix him up properly. He would look great and feel great.

'That's great,' I said faintly, considering the cost of several nights in hospital on a drip with a catheter to his bladder, antibiotics and anti-inflammatories.

Soon he completely recovered. He was himself again. Sleek and shining and huggable, constantly pottering around the house purring. But if the slightest breeze began to stir and rustle the leaves in the trees, Mr Bond would show signs of a new and learnt behaviour pattern. He would scurry through the house, peering out each and every window, his face and whiskers drawn in an anguished display of wretchedness, before hiding under our bed to mope. The storm had dramatically altered his lifestyle. Venturing outside? That clearly was unthinkable: outside was to be avoided at all times. Sleeping on comfy chairs and couches around the house, with the occasional venture out into our leafy courtyard garden, was the only acceptable replacement therapy required for a cat with post-traumatic stress disorder.

In every way now, he was the opposite to his grey and white sister, Miss Moneypenny, with her bold, self-assured personality and dedicated commitment to the removal of mice and rats from the hayshed. Miss Moneypenny would toss Mr Bond a single disgusted look, turn her back and stroll outside to explore our wide-open gardens and farm sheds. Or carefully stalk us as we did our farm chores in the paddocks.

Time went on and the winter cold changed into warm sunny days. Mr Bond began to step out with a new-found confidence and Miss Moneypenny was now clearly impressed. From her favourite place on top of the barn roof she would watch him as he tiptoed through the courtyard gardens, down through the goat paddock to check on the guinea fowl, and then back to the farmhouse. His little jaunts were brief. Sometimes he toured through the goat pens, visiting his favourite places, but he was never outside for long. He was always round about, snoozing on the couch or the beds, or padding just behind us as we busied ourselves in the house.

We grew used to the sight of Mr Bond living in the house. But one afternoon, we spotted him trotting bravely down through the first sheep paddock. He was gone for a while and he didn't come back. We were so worried and searched everywhere, shouting his name. This was somewhat out of character, but as he had become an enthusiast of comfort, I couldn't imagine him having the incentive to travel far.

It was quite late that same evening when the phone rang. It was Shorty and he said he thought our cat was over at his place. Were we missing a male white cat with ginger tips on his ears and tail? The kids had fed it as it was really hungry and Shorty had removed a few ticks from the fur around the cat's nose. I was quite taken aback. It seemed very unlike Mr Bond to have travelled so far – but certainly the description matched his physical appearance. Shorty went on about how his kids had taken to the cat and we got the impression that they were really wanting to keep him.

'Sounds right,' The Engineer said, as he agreed to drive over and pick up Mr Bond. I was relieved but confused as to why our nervous little cat would be in someone else's kitchen. Since Shorty is a near neighbour it wasn't too long before The Engineer arrived back home carrying a male white cat. It seemed the cat liked the ute and had sat quite calmly in the front seat for the drive back. But The Engineer was looking puzzled as he held the cat out to show me.

Chapter Thirteen: When Mr Bond Blew Away!

He did have ginger tips on his ears and tail, and he did look exactly the same as Mr Bond. But we were confused. The problem was, he didn't act like our cat. He didn't purr when I held out my hand to pat him, and when we put him down in the house, he gazed fixedly at us and then looked around with mild interest. After a few moments, I shook my head and turned to The Engineer.

'No, no. This is not Mr Bond,' I whispered. 'It looks like him. Actually, just like him. Identical! But it isn't him.' With a feeling of disbelief, we looked blankly at each other before The Engineer phoned Shorty.

'Mate, I'm afraid this isn't our cat. Sure, it looks the same, but there's no doubt about it. He must be a stray.' There was a pause and I heard The Engineer tell him that this was a bizarre case of mistaken identity. Then he gathered up this cat that was the same as Mr Bond, but not Mr Bond, to return him to Shorty's kids who were eager to keep him.

I was conscious of an increasing despair. This was not our cat so where was the real Mr Bond? I was thoroughly upset and trying hard not to fear the worst. I found it difficult to get my head around the fact that Mr Bond was still missing.

It was much later that evening when we heard a meow out in the darkness, and sure enough, a white cat with ginger tips on his ears and tail was scratching at the door for us to let him in.

'Well! Hello Mr Bond!' The Engineer said, picking him up and cuddling him. And this cat, our friendly Mr Bond, purred loudly and contentedly.

None of this made sense. And looking back, we still find it hard to believe that there were two identical white cats, with identical markings, living in our sparsely housed rural neighbourhood. Was Mr Bond a twin? We had picked up our cat and his sister, Miss Moneypenny, as tiny kittens from the local animal shelter. The kittens had been found abandoned in the bush and it was presumed they were part of a wild community of feral cats. So, a brother? Very possible.

Our Mr Bond never strayed away from home again. I like to think he was as relieved as we were that he was snuggly ensconced at home again. The Mr Bond impersonator stayed the night with Shorty's family, but the very next day disappeared and was never seen again. I could never be quite sure why our Mr Bond wandered off on this particular afternoon. But I always had the feeling that maybe this adventure started when he chanced upon a long-lost brother, and maybe they shared a little time together before parting ways – a kind of a cat family reunion.

FOURTEEN

Is our Farm an Efficient Pod?

We had moved into our run-down farmhouse with its forty acres of poorly fenced hills, gullies and thistles with a sense of nervous excitement. The openness and freedom of our new life was exhilarating but our family was still concerned as they asked us the critical question, 'Ma, how long do you think you're going to be able to manage here?'

I was optimistic – we were going to be just fine.

It wasn't long before our days were filled with routine farm duties and figuring out what works best on a farm. There's always time-consuming stuff to do: jobs that have to be done for the animals, for the land and for us. The Engineer lost no time in hiring *Sir Lop-a-Lot* – two energetic, hardworking young men, armed with chainsaws and whipper snippers. They lopped and chopped and trimmed until the 'before and after' differences were simply staggering.

Fencing

It didn't take The Engineer long to organise fencing contractors to construct reliable, well-designed fences to prevent our animals straying and upsetting neighbours. The original farm fencing was undeniably a useless frame around the grazing land, with rotten wooden corner posts, wire draped artistically around any tree, and gaps big enough to run the tractor through.

The fences over our gullies have been well pegged down, as pickets can lift in drought conditions and thrust fences upwards – leaving surprisingly big gaps. And after we needed to rescue a pregnant ewe trapped in mud up to her belly in one of our dams, we fenced this dam off as well.

All our paddocks have names, although not the most original: bottom paddock, middle paddock, goat paddock, side paddock, night paddock and new paddock. There is even a Belle paddock. The night paddock has an electric fence around the sheep yards, barns and adjoining small paddock to protect our sheep and goats from feral and free-roaming domestic dogs during the hours of darkness. It wasn't long after we arrived that we heard stories about feral dogs roaming down from the hills into our valleys and beyond. Occasionally, we have heard wild howling in the evenings – eerie (to say the least), lonely, mournful cries floating across the dark edges of distant hills, sending shivers down my spine. So, we run our German Shepherd dogs through the paddocks every morning before the sheep and goats are let out. With their sniffing, peeing and frequent poops along the boundary fence our dogs assert domination over their territory.

Weeds

One of the toughest problems for paddock management is getting rid of weeds. Our sheep keep the hillsides looking amazing with short-cropped grass, but short-cropped grass encourages weed growth, and eradication becomes a long-term, never-ever-ending field project. Paying attention to weeds was something I was to learn, because these little plants can grow huge after one tiny drop of rain, especially lantana and chinee apple.

And yellow fireweed! They're like daisies. If you have one in your paddock, it looks pretty and unique. If you fail to root it out, however, you find five the next day … fifty after that … and then your farm is totally, completely and profligately covered with fireweed. By then you see them for the weeds they really are.

We tried spraying weeds – a job we dislike and ineffective in large paddocks. And our hills and slopes are too steep for slashing weeds. Then Aunty Tup said something interesting. 'Get yourself some Boer goats! They're famous for eating through weedy paddocks.'

And they have. These practical, sturdy animals have transformed our hillsides by happily munching their way through woody weed, fireweed and other undesirables. All except thistles.

'Why not thistles?' I asked the goats one day. 'Even Eeyore* eats thistles.'

Because our farm is pint-size and hilly, we have managed to get away with simple methods to remove the thistles our goats won't eat. One is by smashing them – the thistle dies quickly when the centre is damaged. However, after rain The Engineer gets busy with our brush cutter and blade. This is fast, furious and most effective.

How to Do Wonders with a Little Land

Someone once asked me why we bother with a garden when we only have access to tank water. My answer is simple: I have a passion for gardens.

After we moved onto the farm, we made the decision to renovate and paint the cottage and plant a bush garden with trees and real lawns, where once there was nothing but rubble, stones and weeds. Our garden is loosely designed so it is easy to maintain and as every enthusiastic gardener knows, sheep poo and horse droppings feed the soil and the plants, while lots of poopy hay mulch keeps water use to a minimum.

There's something very satisfying about growing our own fruit and vegetables, and with the luxury of a huge backyard, we have planted our first real orange tree. Every year, we plant cherry tomatoes and strawberries full of sweet flavour, and there's nothing nicer than eating these straight from the plant.

It wasn't long before we made new country-style friends, sharing enthusiasms and taking pleasure in bringing each other things like jam, bougainvillea and silky oak plants. Bougainvillea is considered by some to be a disaster but it's hardy, which means short of digging it up, it will cope with being neglected and still manage to flourish with amazing colours.

Birds and wildlife love gardens. We have bees and butterflies in our flowers, and tiny wrens showing flashes of red or turquoise as they meet inside our hibiscus hedge. Owls perch on the farm gates at night, blinking with huge eyes, and we cope with noisy plovers screeching and screaming in the grasses.

And trees? We have planted lots of them in strategic positions. With the trees have come the birds. Laughing kookaburras, lorikeets, yellow and blue budgies, a choir of butcher birds and lots of magpies. And I'm hoping that the type of she-oak tree I've planted is the one that the black cockatoos feed on when the seeds pop in the cones.

On the other hand, we also have lots of things that can kill in a nasty way. I've even had a rather scary but unique experience of a snake slithering between my legs. And I could not help but notice when bull ants swarmed over my shoes and threatened to invade my jeans. We have a few wallabies but fortunately, no kangaroos. Kangaroos are wild animals that can grow huge, have the capability of kicking and punching, and no matter how cute they seem to be, have been known to entice domestic dogs into dams, holding them under water until they drown.

The weather is a constant challenge and shady trees are a must in the heat of our sweltering summers. If there is no rain, which is so often the case, the season is punishing for any grassy paddock. The heat can be so intense that we need to hang shade covers and tarpaulins over animal water containers to prevent the water overheating, which can also lead to the build-up of algae growth.

After the summer sun has slammed down and scorched us every time we venture outdoors, there's nothing in the world like teeming rain. On our muggiest and hottest spring and summer days, the sky begins to grey over and fill with dark clouds. And the storms are spectacular – crashing thunder and wild flashes of lightning as the wind groans and moans. Big fat raindrops soak into the parched fields and afterwards the smell of rain lingers in the hot night air as the shriek of the cicadas intensifies. We notice the transformation in our landscape

Chapter Fourteen: Is Our Farm an Efficient Pod?

almost at once, as water begins to trickle down the slopes, tumbling through and over the rocks and gullies, creating tiny rushing streams, like silver ribbons, moving into dry dams. Frogs take their cue to begin clicking and belching, and toads croak to each other across the new expanse of water.

Sheep are supposed to be adaptable, taking advantage of surrounding terrain such as hills, ridges, trees and shrubs for shelter. Their wool is a good insulator. But, let a little rain fall and there's a mad dash as the sheep and goats head for the barns and undercover yards with an urgent call to young lambs and kids to get inside.

In the extreme cold of winter, the weather starts to close in and animal shelters and pens need to be waterproof, draught-proof and virtually indestructible. The Engineer has built little roofed sheds out of pine and pickets, and practical shelters out of water tanks. Each shelter has a wooden floor and opens out to grassy inter-connected pens. Lining the lambing pens with straw bales makes for cosy sheltered areas where I can isolate the needy, nursing, or frail sheep and feed them special meals of grain and molasses for extra calories and protein.

We need to keep tiny lambs warm. There is that option to put tiny coats on the lambs, like small dog coats. How cool is that? I have to say, I haven't gone this far. And if I did, I couldn't bring myself to tell Clay! Or Scooter!

The only downside to having sheep and goat housing is having to continuously scoop up poop. A clean-up task precious few handle with enthusiasm. After wet nights the sheep and goat shelters will be an awful disgusting mess. As in blllerrrrgh! So, in response to my groans, The Engineer built easy-to-clean wooden floors to sweep and hose and rake. And the best part – poopy hay for my garden.

Keeping Abreast of Farm Stuff

1. Feet

Our sheep and goats graze over hilly, stony paddocks and rarely have issues with lameness or hoof problems; and our mostly dry climate means we have never experienced the issues of contagious foot rot. In springtime, Wally trims the hooves of all sheep and goats as he shears the sheep.

2. Pink Spray

Purple or Pink Spray is an antibacterial wound aerosol and fly repellent for use on open wounds and is our first response to managing lumps, bumps, cuts and chewy tails. We make sure we always have cans on hand.

3. Feed with benefits

Although we do have reasonable paddock grass in most seasons, sooner or later, the hot dry temperatures come about. If no rain falls, the paddocks begin drying off by the minute and, before we know it, we're dishing out lucerne hay. And NOTHING stands in the way of a mob of sheep and their daily serves of lucerne. They gobble it up like we eat hot fudge.

As well as supplementing with lucerne, we also feed our animals grassy hay, round bales, barley hay, oaten hay, and other hay and grasses with fancy names and extra special additives. A chickpea and lucerne combination is seasonal, and fine for sheep and goats to browse on. The sheep are not so keen on the grassy hay choices and display a rather disappointed attitude as they check out each wisp before reluctantly beginning to chew.

To quote a distant neighbour, 'They'll eat it if they're hungry enough.' But his laissez-faire argument is not persuasive, as in my experience, if they don't like it, they will shit in it, pee over it, trample through it – and leave a discarded mess for my garden.

Careful planning is required when dishing out feed. The cows are easy because we can easily throw biscuits of hay over the fence, but the sheep are a more complex operation. The sound of the tractor in action triggers a stampede up the hill and once, I was tossed to the ground by these huge woolly powder-puffs on legs and unceremoniously trampled on as my legs buckled and folded beneath me. We have since learnt that the sheep can smell hay from a distance, and will charge up the hill at top speed and through the gates in a solid block. So, it's best to put the hay out *before* we open the gate.

One morning, The Engineer left the gate open and had his legs taken right out from under him. He lay on the ground stunned, gasping for breath, as I staggered around laughing helplessly. With an effort, I helped him up and managed to brush most of the hay off his face and shoulders. It was not appropriate to say *'I told you not to …'* as he was in no mood to appreciate the funny side of being bowled over, trampled on and pushed aside.

The Bottom Line Regarding The National Livestock Identification System (NLIS)

One of the most common methods of identifying livestock visually or electronically is to apply an ear tag. In Queensland, choosing the most appropriate ear tag to use is governed, in part, by regulations. Ear tags in metal or plastic vary in size, colour and shape. They are made by a number of companies, each with their own design.

Sheep and goats are good at catching their heads in fencing wire and we have had some minor ear damage, so it was important we made

HARRI-HENRY'S FARM

sure we use the appropriately sized tags for sheep, goats and cows. Manufacturers do provide application instructions, which are important to follow, to ensure tags stay on and to avoid ear injury.

All cattle must have an electronic national livestock identification tag (known as NLIS tags) before being moved between properties. And all sheep must have a visually readable identification tag.

Every property with livestock must have a Property Identification Code (PIC), and both cattle and sheep tags must be stamped with it. This applies to anyone with cattle, sheep, pigs or goats.

So many lists of do's and don'ts! The available information is full of big words, complicated strategies and techniques, and just keeps growing and changing – nothing ever seems easy. To learn more about it, I phoned our local Department of Primary Industries.

To obtain a PIC we needed to make an application for the registration of our property. A PIC was assigned and remains permanently with the property. This spatially defines property where animals live, and is used to identify property on movement document forms, vendor declarations and on the NLIS database. The PIC is printed on the NLIS tags.

Chapter Fourteen: Is Our Farm an Efficient Pod?

The NLIS is a system that identifies animals so that they can be traced. A key benefit is emergency disease control. Compulsory identification of animals also helps reduce the risk of livestock being stolen and increases the chances of livestock being returned, if they stray or are stolen.

Livestock rustling is a problem and increasingly, a booming business. Every so often our local newspaper has stories about cattle theft and the Queensland Police Stock Squad shares warnings and precautions for all property owners. Calves are easily stolen. Nets are usually thrown over the calves and they can be trapped and removed without the animals making any disturbing sounds.

Lockman told us about his own experience with cattle rustling. Early one morning, he noticed his locked gates were showing signs of forced entry so he contacted the police. The Stock Squad promptly turned up soon after and arrested a gang of armed thieves with large dogs, rounding up Lockman's cattle into trucks.

A new high-tech method of tracking stolen sheep is being field-tested in England. This is a biological branding technique which can't be altered or removed by thieves. A liquid containing a unique DNA code is applied topically. It can't be detected by the naked eye, is non-toxic, and can last up to five years. If needed, law-enforcement officials can use an ultra-violet lighting device to detect the presence of the product and cross-reference it with a database. Notices around the farm will hopefully deter thieves, but where livestock are stolen, the markings will trace animals back to the farm and enable spot checks at places like auction marts, stock auctioneers and meat works.

Eeyore is a character in The Complete Winnie The Pooh *by A A Milne*

Sources:

www.fwi.co.uk/news/crime/uk-farmers-see-sharp-rise-in-sheep-rustling

www.fwi.co.uk/news/crime/technology-helping-farmers-curb-sheep-thefts

FIFTEEN

The Finish Line

So, What's it Like to be a Farmer?

There is truth in the old cliché, 'Nothing ventured, nothing gained'. And ever since we grabbed the opportunity to buy a farm, The Engineer and I have proved that we can make a life for ourselves on the land. Farming animals has been more fascinating than I ever would have guessed and given us such a new sense of purpose.

Our farm lies in a cattle and sheep district, and one thing that's great about this rural community is the support and encouragement we have enjoyed. We have met so many generous, warm people and we feel grateful for the friendliness they always show us.

All except for Bob, that is.

The most awkward experience that sticks in my mind is the time I was challenged by Bob. Bob is the owner of a western Queensland mixed cropping, cattle and sheep property, and is very self-opinionated. I'd learnt that no one seemed to like Bob, or even his wife, Maud.

'He's as popular as a turd in a swimming pool,' Dusty whispered in my ear when we saw Bob in the distance one day. 'If he can say a hundred words instead of ten, he will do so, and he tells cock-and-bull stories which no one believes.'

One morning I came upon Bob outside the local produce store. He was walking briskly past the shop entrance and he stopped short

Chapter Fifteen: The Finish Line

when he saw me. Despite my cheery hello, it was clear from the start he disliked my forthright manner. I knew he was assessing me, taking in my senior person's appearance and presuming on my inexperience. But I was entirely unprepared for him to look me up and down in a discouraging way, shrug his shoulders and say, 'City folk know nothing about farming and farming's not for old women!'

He was brisk and civilised, but I felt a good deal affronted by this and my face reddened as he went on.

'You know the difference between you and me?' He was loud and all up in my face. 'I farm for a living. But you – you're just a hobby farmer with your menagerie of hand-reared pet animals.' He gave me a sour look.

I stared blankly up at him. I had been trying so hard to be a serious and awesome farmer, and I thought that overall I was doing a pretty good job. I had been loving the satisfied feelings that I'd recently been experiencing, telling me that I had really done something good with our retirement – making me feel I'd been waiting for this all my life.

I'd had enough. I made a dismissive gesture as I looked Bob straight in the eye with a hostile stare and said, 'Get stuffed!'

But his words were still spinning in my head and I was stunned by what he had implied. Not just by his severe tone, but by the implication that all hobby farmers are a pain in the arse. I felt miserable and stupid as I remembered some of the really dumb things I'd done. So, what made me think I was getting good at farming?

One night not long after this chance meeting with Bob, The Engineer and I were eating hamburgers out in the café gardens. I was enjoying the gossip and the chitchat, and then I asked Scooter, Eddie and Earthy Murphy if they viewed hobby farmers like me, with my enjoyable mix of animals, as a pain in the arse.

Eddie paused, thinking it over very carefully, and then said, 'So why are we asking?'

I groaned inside and told them how I'd had words with Bob.

Eddie scowled. 'He's got tickets on himself big as posters, mate. Flamin' posters! Thinks the sun shines out of his bum!'

As for Scooter – he leaned back in his chair and laughed and laughed as he declared that hindsight permitted him to suggest that Bob smelt like a goat's underpants!

'Maaaaaate!' Earthy Murphy leaned over and thumped the table. 'You want to know one thing that's great about you and Engineer Man? Your thing – renovating and working Clifftop Cottage Farm – is one of our district's greatest achievements. Hey, I thought you old dudes were never going to be able to pull it off.'

'I never thought you'd do it,' Eddie mumbled in disbelief. 'You showed us you could kick everyone's arses.'

'Ah, the point is, my dear, that being a farmer is not about the size of your property.' Aunty Tup leaned forward and put both hands on the table as she butted into the conversation. 'I can tell you that it's not about how long you've been farming. It's about you and your animals and whether you can muddle through the really tough shit.'

Hmmm! I couldn't help but feel a sense of pride as they went on to joke about us. The weird retired townies learning to run a farm. Especially one on top of a range of steep hills. Somewhat to our surprise they made us feel accomplished and clever, even if their comments seemed too good to be true.

The Engineer and I *are* getting fitter and stronger, and feeling more alive as life at Clifftop Cottage Farm just keeps getting better, more fascinating and more refreshing. A magic fit for our retirement.

About the Author

A teacher, jazz musician and award-winning children's music storybook writer, Chrissy Tetley is a mum of two boys. She lives with her retired engineer husband on a small sheep and cattle farm in the high blue hills of Queensland, Australia.

Chrissy's love of story telling led her to write the *Music on the Bookshelf* series of children's books. Featuring Australian animals, the popular series aims to subtly educate children about music. These books were named among the best in the US Mom's Choice Awards that honour excellence in family-friendly media, products and services, where they achieved one gold and three silver awards.

When not writing, Chrissy plays the flute in a swing jazz band and she has won Grand Champion awards in obedience with her German Shepherd dogs.

About the Illustrator

Gary Young lives in Auckland, New Zealand where he works as a freelance illustrator and cartoonist. His work includes cartoons and illustrations for magazines, newspapers, children's books and commercial work. For 15 years, he worked as an artist for an Auckland publisher on advertising work and creating illustrations and cartoons for their magazines and newspapers. Since then he has freelanced, working for clients, first in New Zealand then England and more recently for 7 years in Brisbane Australia before moving back to his home country.

Gary specialises in giving a visual dimension to characters and scenarios or adding human characteristics to animals or inanimate objects.

www.ingramcontent.com/pod-product-compliance
Lightning Source LLC
Chambersburg PA
CBHW040241010526
44107CB00065B/2826